T0210580

Routledge Explorations in Energy Studies

Decarbonising Electricity Made Simple
Andrew F. Crossland

Wind and Solar Energy Transition in China
Marius Korsnes

Sustainable Energy Education in the Arctic
The Role of Higher Education
Gisele M. Arruda

Electricity and Energy Transition in Nigeria
Norbert Edomah

Renewable Energy Uptake in Urban Latin America
Sustainable Technology in Mexico and Brazil
Alexandra Mallett

Energy Cooperation in South Asia
Utilizing Natural Resources for Peace and Sustainable Development
Mirza Sadaqat Huda

Perspectives on Energy Poverty in Post-Communist Europe
Edited by George Jiglau, Anca Sinea, Ute Dubois and Philipp Biermann

Dilemmas of Energy Transitions in the Global South
Balancing Urgency and Justice
Edited by Ankit Kumar, Johanna Höffken and Auke Pols

Assembling Petroleum Production and Climate Change in Ecuador and Norway
Elisabeth Marta Tómmerbakk

For more information about this series, please visit: www.routledge.com/
Routledge-Explorations-in-Energy-Studies/book-series/REENS

Assembling Petroleum Production and Climate Change in Ecuador and Norway

Elisabeth Marta Tómmerbakk

Routledge
Taylor & Francis Group

LONDON AND NEW YORK

First published 2022
by Routledge
2 Park Square, Milton Park, Abingdon, Oxon OX14 4RN

and by Routledge
605 Third Avenue, New York, NY 10158

Routledge is an imprint of the Taylor & Francis Group, an informa business

© 2022 Elisabeth Marta Tómmerbakk

The right of Elisabeth Marta Tómmerbakk to be identified as author of
this work has been asserted by her in accordance with sections 77 and 78
of the Copyright, Designs and Patents Act 1988.

All rights reserved. No part of this book may be reprinted or reproduced
or utilised in any form or by any electronic, mechanical, or other means,
now known or hereafter invented, including photocopying and recording,
or in any information storage or retrieval system, without permission in
writing from the publishers.

Trademark notice: Product or corporate names may be trademarks or
registered trademarks, and are used only for identification and explanation
without intent to infringe.

British Library Cataloguing-in-Publication Data
A catalogue record for this book is available from the British Library

Library of Congress Cataloging-in-Publication Data
Names: Tómmerbakk, Elisabeth Marta, author.
Title: Assembling petroleum production and climate change in Ecuador
and Norway/Elisabeth Marta Tómmerbakk.
Description: Abingdon, Oxon; New York, NY: Routledge, 2022. |
Series: Routledge explorations in energy studies | Includes bibliographical
references and index.
Identifiers: LCCN 2021006922 (print) | LCCN 2021006923 (ebook)
Subjects: LCSH: Climate change mitigation–Political aspects. |
Petroleum industry and trade–Political aspects–Ecuador. | Petroleum
industry and trade–Political aspects–Norway. | Environmental policy.
Classification: LCC TD171.75 .T66 2022 (print) | LCC TD171.75
(ebook) |
DDC 363.738/74–dc23
LC record available at https://lccn.loc.gov/2021006922
LC ebook record available at https://lccn.loc.gov/2021006923

ISBN: 978-0-367-60780-7 (hbk)
ISBN: 978-1-032-04828-4 (pbk)
ISBN: 978-0-367-60781-4 (ebk)

Typeset in Baskerville
by Deanta Global Publishing Services, Chennai, India

To my dear father

Contents

Figures

Acknowledgments

The greatest adventures in life are often those experiences we collectively build and share with others. This research project has definitively been an adventure, thanks to many people who have contributed, participated, oriented, and guided the process. I want to express my gratitude to all my informants in Ecuador and Norway for their generosity and willingness to share information and insights. Without their "voices", this book would simply not exist.

I am enormously thankful to my doctoral supervisor Professor Randi Kaarhus, for strongly believing in this research project and the possibility of reaching a broader audience. Her continuous support, advice, and professional experience has been invaluable. I simply cannot thank her enough for her commitment and guidance. My sincere thanks also go to Dr. María Guzmán Gallegos, my co-supervisor, for her comments, reflections, and insights on Amazonia, which have been of great importance for the project's development. Furthermore, I wish to thank the Faculty of Social Sciences at Nord University in Norway for granting me the PhD fellowship that made this inquiry possible. A special thanks to all the members of the Research Group for the Environment, Resource Management, and Climate, led by Professor Berit Skorstad. Furthermore, I owe thanks to the authorities at the University of Cuenca for their support as well as Professor Kelly Swing from the University of San Francisco de Quito, for helping out with the necessary arrangements for my field trip to Yasuní. A special thanks to Annabelle Harris and Matthew Shobbrook at Routledge for guiding me through the process. Working with you has been a true privilege.

Last but not least, I want to thank my wonderful parents and family for their continuous support, prayers, and encouragement. Thank you for always reminding me that the purpose of this book is to share and multiply the knowledge acquired together with others in future collective dreams and projects. A special thanks to my sister María and my dear husband Alberto for being who you are and for helping out with big and small matters during difficult times for the family so that I could continue writing. Without you everything would have been much more difficult. This effort is definitely also yours.

Abbreviations

ANT	Actor-Network Theory
API	American Petroleum Institute
CCS	Carbon Capture and Storage
CGY	Certificados de Garantía Yasuní
CDES	Centro de Derechos Económicos y Sociales
CDM	Clean Development Mechanism
CEPAL	Comisión Económica para América Latina y el Caribe
CEPE	Corporación Estatal Petrolera Ecuatoriana
CONAIE	Confederación de Nacionalidades Indígenas del Ecuador
COP	Conference of the Parties
DRA	Drag Reducing Agents
ECLAC	Economic Commission for Latin America and the Caribbean
GDP	Gross Domestic Product
GHG	Greenhouse Gas
IEA	International Energy Agency
ILO	International Labour Organization
IMF	International Monetary Fund
IPCC	Intergovernmental Panel on Climate Change
ITT	Ishpingo, Tambococha, and Tiputini
LOVESE	Lofoten, Vesterålen, and Senja
NCS	Norwegian Continental Shelf
NGO	Non-governmental Organization
NICFI	Norway's International Climate and Forest Initiative
NORAD	Norwegian Agency for Development Cooperation
OECD	Organization for Economic Co-operation and Development
OFD	Oil for Development
OPEC	Organization of the Petroleum Exporting Countries
REDD	Reducing Emissions from Deforestation and Forest Degradation
SIL	Summer Institute of Linguistics

SOTE	Sistema de Oleoducto Transecuatoriano
UNDP	United Nations Development Programme
UNFCCC	United Nations Framework Convention on Climate Change
WTI	West Texas Intermediate
WWF	World Wildlife Fund
ZITT	Zona Intangible Tagaeri Taromenane

1 Introduction

Presenting the problem: The fossil fuels–climate change nexus

Today, a widespread agreement exists within the scientific community regarding anthropogenic global warming (Chevalier 2009a; Bradshaw 2014; McGlade and Ekins 2014). There is little doubt, in other words, that climate change will constitute the greatest environmental challenge in the years to come. The gradual increase in temperature will most certainly have unprecedented consequences not only for the environment but also for the individual and for society as a whole. Therefore, in recent years, climate change has gone from being a topic among researchers and experts to an important political challenge that requires governance at many levels: International, national, and regional (European Commission 2007). Today, climate change is at the heart of environmental discussions and is increasingly replacing ecology, environment, and sustainable development, as the overarching theme (Lidskog and Sundqvist 2013: 19) when discussing the impacts and implications of human-nature interactions.

Global warming is largely a consequence of the man-made global energy system currently dominated by fossil fuels. The possibility of limiting its impacts in the future "will depend critically on whether energy use can be greatly reduced, dissociated from carbon or both" (European Commission 2007: 13). Accordingly, climate change can be conceived as a new kind of energy crisis that clearly differs from previous conceptions related to scarcity, that is, energy constraints due to high oil prices or depletion of oil and gas reserves (Chevalier 2009a: 1). In the past, many policymakers working with energy security issues were largely concerned with different scenarios and challenges tied to conceptions of "peak oil". The focus was placed on when oil production was estimated to peak and decline. Accessing new oil reserves was, therefore, considered vital to secure stable energy supplies. Today, these concerns are considered less pressing due to technological innovation, which has improved the resource recovery rate considerably (Chevalier 2009b; Bradshaw 2014). While factors that influence the possibility of accessing particular petroleum resources include both technical and economic viability in combination with geopolitical aspects, there is also an expressed need to take into account possible future "reductions to oil availability arising from constraints placed on [carbon dioxide] CO_2 emissions" (McGlade and Ekins 2014: 103). Hence, today's energy crisis is not directly related to whether the world is approaching peak oil or not, but to the fact that our energy system is no longer sustainable in a 2°C scenario established as a political goal by the United Nations Framework Convention on Climate Change (UNFCCC). From this perspective, the current energy crisis is not primarily about price volatility or exhaustion of petroleum resources but the excess of hydrocarbon reserves available for consumption when the political aim is to maintain the global rise in temperature below 2°C. Hence, "sustaining fossil-fuel production will come at a cost, both economically and environmentally" (Bradshaw 2014: 28). While the exact costs are impossible to calculate, the economic impact of climate mitigation could be high, especially in developing countries, if efficient measures are not agreed on internationally and implemented within a short timeframe. Ever since the international climate negotiations started in 1992 with the Rio Summit, there has been an ongoing debate regarding what constitutes efficient strategies to reverse the gradual temperature rise and how to distribute the costs of mitigation globally. In other words, how to share the burden between countries. This brings us to another related issue, namely, the global imbalance between developed and developing countries when it comes to access and consumption of energy resources. While 18% of the world's population is responsible for close to 50% of the world's energy consumption and around 30% of all greenhouse gas (GHG) emissions, many people around the world do not have access to basic services such as electricity or clean water, necessary conditions to achieve economic development and welfare (Chevalier 2009a: 1).

Global warming produced by changes in the chemical composition of the atmosphere is closely related to two important processes: Industrialization and urbanization. Since the Industrial Revolution started in the West, the concentration of GHGs has increased considerably, especially carbon dioxide, which has risen by 30% since 1880 (Giddens and Sutton 2013: 177). From the view of economics, higher levels of GHGs in the atmosphere are conceived as externalities. In *The Stern Review*, climate change is described as "the greatest example of market failure we have ever seen" (Stern 2007: 1). This economic approach towards climate change is also the logic that underpinned the Kyoto Protocol: Through market mechanisms, these externalities could be handled and economically accounted for by turning GHG emissions into a tradable commodity in a market designed for this purpose.

According to climate scientists, higher levels of CO_2 in the atmosphere are mainly caused by the burning of fossil fuels, but industrial production, deforestation, large-scale agriculture, and vehicle emissions also play important roles (Giddens and Sutton 2013: 177). In this context, the International Energy Agency (IEA) has stated that in the absence of large-scale implementation of carbon capture and storage technology (CCS), a considerable part of the world's known fossil-fuel reserves must be kept in the ground so as not to surpass the 2°C goal (IEA 2012). Similar carbon budgets have been put forward by several environmental think tanks and climate organizations. However, the thesis of "keeping fossil fuels in the ground" (Princen et al. 2015: 7) has not really gained momentum.

Scope and approach

This book addresses some of the controversies and uncertainties that arise from the extensive exploitation and use of fossil fuels and their role in global warming. The scope is not to discuss climate science per se but its implications for political decision-making and the construction of national policies that enable reconciling multiple and often conflicting agendas. Emphasis is therefore placed on how oil and climate change are "entangled" and how this relationship is politically handled and worked upon at different levels. Important points of departure are the various environmental, political, economic, and technological challenges related to climate change and the required transition towards a low-carbon energy system. The book specifically explores the question of why a transition towards a "post-carbon" society is so difficult to achieve by examining how the relationship between petroleum production and climate change is politically framed and negotiated in contested cases. The question is approached through a comparative case study of Lofoten and Yasuní-ITT (Ishpingo, Tambococha, and Tiputini), located in Norway and Ecuador, respectively. The cases constitute relevant sites to empirically study how the relationship between oil and climate change is framed and enacted as they both belong to oil-exporting countries with highly oil-dependent economies and therefore have a strong stake in the issue. It is important, however, to point out that oil exploitation in Norway and Ecuador takes place under very different technological conditions and in very different environments. Norway's oil production is offshore, which means that petroleum activities do not interfere directly with people's lives, neither are possible environmental risks tangible in the same way as in Ecuador, where oil extraction is carried out in the Amazonian rainforest. Here, local inhabitants and communities live in close proximity with national and international oil companies and have direct contact with petroleum installations and infrastructure. As a consequence, environmental risk is often directly part of their everyday lives and experiences. This situation has developed into strong opposition and processes of resistance among the local population. In the case of Norway, the condition of operating offshore may possibly explain why the oil industry does not face similar opposition. The petroleum debate in Lofoten, however, is different, as it has a strong local component due to the importance of the fisheries and tourism for the area. Here, contrary to what is the case with offshore petroleum operations in the North Sea, many local inhabitants oppose the possibility of opening up the area for oil activity, as they fear that the incursion of the oil industry will generate territorial disputes with the local fisheries and change the image that outsiders, especially tourists, have of Lofoten. The local opposition is based on concerns regarding the risk petroleum production potentially represents but also on identity matters (see Kristoffersen and Dale 2014).

Despite being two very different cases with different technological, environmental, and political conditions for oil activities, the cases also have certain things in common. Selecting two cases from different parts of the world (one from a developed country and another one from a developing country) enables me to examine the research problem from local, national, and global perspectives: In other words, to trace links and connections across borders and scales. However, it is important to emphasize that the cases are not the geographic places themselves but the ongoing political processes and controversies related to the question of extracting oil or not in these environmentally sensitive areas. I also want to emphasize that the scope of this inquiry is not a holistic analysis of the cases, as there are many aspects related to the Lofoten and Yasuní-ITT controversies that I do not investigate.

While the case study literature frequently distinguishes between "holistic" and "embedded" case studies, the meaning of holism seems to vary considerably. According to Bartlett and Vavrus, understandings of holism tend to be based on "a traditional notion of culture and a functionalist theoretical stance" (2017: 37). In the past, ethnographic studies were often concerned with describing a whole culture or way of life, a practice that tended to be underpinned by a bounded understanding of culture (Bartlett and Vavrus 2017: 37), which complicates examining how boundaries are "enacted into being" (Law 2007: 12) in the first place besides diverting our attention away from more processual understandings of the world. Bartlett and Vavrus propose "an iterative and contingent tracing of relevant factors, actors, and features" (2017: 37) to compensate for these methodological shortcomings. In line with their perspective, the aim of this book is not to give a comprehensive or complete account of all the components that comprise the cases but to get close up on certain links, connections, and relationships to understand how they are politically, environmentally, and economically produced and made viable across levels and sites. I chose to work with a science and technology studies (STS)/actor-network theory (ANT) approach to fulfill this purpose, mainly drawing on authors such as Latour, Law, Asdal, and Callon. This framework/ method will be further explained in the next section. The idea has been to follow the cases as unfolding processes by focusing on the various ways in which actors, entities, objects, science, and policies are interweaved and circulate. I worked with the following subsidiary research questions to guide the empirical analysis:

1. How are different kinds of knowledge, values, discourses, and policy practices produced and articulated in the two cases?
2. How do stakeholders frame and negotiate environmental risk?

Consequently, this inquiry focuses on different challenges and political strategies related to petroleum production and non-extraction policies in connection with climate change and national fossil-fuel dependency. The aim is specifically to examine how different local and national entities and stakeholders link challenges related to petroleum activity and climate change. Moreover, in our globalized world, there seems to be a contradiction between the desire to guarantee economic growth and development, especially in developing countries, and at the same time preserve the environment, biodiversity, local culture, livelihoods, and indigenous peoples' territorial rights. These latter concerns, in particular, make it important to try to discover how certain values, knowledge systems, discourses, and policy practices are produced, circulate, and become legitimate within the petroleum debate.

Actor-network theory

ANT is associated with the work of Bruno Latour, John Law, and Michel Callon. While sometimes referred to as the Paris School of Science and Technology Studies (Hess 1997), an important difference between ANT and other perspectives within STS is the blurring of the ontological divide between human and nonhuman actors. This issue has caused heated debate, especially between advocates of the Strong Programme developed by Bloor (1976) and theorists working within an ANT perspective, as

> ANT tends to be seen as an attempt to erode, or at least "bypass", the barriers between the natural and the social arena. In contrast, through its maintenance of a subject-object distinction, the Strong Programme is sometimes portrayed as protecting those barriers, especially by ANT writers.
>
> (Newton 2007: 28)

This controversy takes us to the core of ANT, that is, how humans and nonhumans are connected and become associates through extended networks. Actor-network theorists often prefer the term "*actant*" instead of "actor", since "actor" in social sciences is a concept heavily loaded with meanings that vary from the purposive and calculating actor of rational choice theory to the actor guided by social norms and values in Parsons's structural functionalism. Hence, from a traditional, socio-logical perspective, the concept of the actor refers to human beings with inten-tionality and objectives, which consequently motivate their actions. Alternatively, the use of actant, which has its origin in semiotics, is revealing because it illus-trates that the actors are not important in themselves but are activated and gain their significance in relation to other entities. Consequently, "ANT adheres to a material, expanded version of semiotics, also called relationism or associationism, which studies how things come into existence as a result of the set of relations of which they are a part" (Asdal et al. 2007: 29). Agency becomes a matter of attri-bution, as the entities' ability to act is not an inherent characteristic but depends on their position or location in the network (Hess 1997: 108). From an ANT per-spective, sociology's definition of the social as an exclusively human possibility has generated a reductionist view of both agency and interaction.

What differentiates ANT from previous approaches in STS is the extended applicability of symmetry, as material entities also play a significant role in producing relational effects or outcomes. Law (1987) refers to this process as "heterogeneous engineering". In other words, this is how a variety of elements and entities with different origins and locations are assembled in and through sociotechnical processes. Facts and technologies circulate and associate with other entities as their networks expand and, thereby, become more robust and reliable. Through translation networks, inscriptions, and inscription devices, technologies and human actors are brought together and engage in interaction (Callon 1995: 52). When an increasing number of actors and actants are enrolled in the network, it becomes more powerful and less questionable. This is the reason ANT is referred to as a sociology of translation (Callon 1986), emphasizing the relational dimension as to what causes the effects and outcomes. According to Callon (1986), there are four moments in the translation process: problematization, interessement, enrollment, and mobilization of allies. Problematization refers to how a problem is defined and how one can persuade others to accept this problem definition in order to become an obligatory passage point. Interessement is related to a series of actions that impose and stabilize the roles and identities of other actors defined through problematization. Enrollment follows successful interessement, as actors accept the interrelated roles that have been designed and attributed to them. Finally, mobilization has to do with achieving representation by particular spokespersons (Callon 1986).

Nature, politics, and science intertwined

One of ANT's main objectives has been to overcome a series of dichotomies that underpin and guide common understandings of society, nature, and science – facts/values, culture/nature, subject/object, and politics/science – are all productive *hybrids* of the modern world. Latour (1993) argues that this hidden *hybridization* is part of the Modern Project, which has separated humans and culture in one ontological zone and nonhumans and nature in another one. Latour calls this process "purification" (Latour 1993). Translation is seen as a parallel process that generates heterogeneous mixtures that combine elements and entities, in other words, "hybrids of nature and culture" (Latour 1993: 10). Latour maintains that we find the origin of two coexisting, separate worlds (the social and the natural) in circumstances that occurred during the period of the Scientific Revolution. An important element in this analysis is Shapin and Schaffer's (1985) *Leviathan and the Air-Pump: Hobbes, Boyle, and the Experimental Life*. The book describes a prolonged dispute in the middle of the 17th century between Robert Boyle and Thomas Hobbes, who were both political and natural philosophers. The outcome of their disagreement, however, was that Boyle ended up being recognized as a natural scientist, while Hobbes is regarded as a political philosopher (Blok and Jensen 2011: 56). Consequently, their story is closely related to the understanding of science and politics as two separate and distinct domains of society. According to Latour, the nature/culture divide is the outcome of a combination of ideas, apparatuses, and procedures that Boyle and Hobbes used to settle their controversies that involved "the distribution of scientific and political power" (Latour 1993: 15). From this perspective, the conflict between them is of particular interest, as it constitutes the origin of modernity and what Latour calls the "modern constitution" for which Boyle and Hobbes play important roles as the "founding fathers" (1993: 28). As seen by Latour, the controversy is primarily about how to establish authoritative knowledge and, thereby, which knowledge systems and corresponding procedures should be considered valid. He further maintains that political and scientific representations co-constitute each other, even if modern discourse and practice constantly struggle to keep them apart.

Ordering and framing multiple realities

Overall, the material-semiotic identity of ANT has become increasingly important in later developments as it points towards networks as uncertain, precarious, and open-ended processes in which actors and entities and their size and power are relational effects of their mutual encounters and the specific position they hold in the network. This anti-essentialism can also be traced back to Foucault's conception of discursive practices and their physical and material consequences for reality. Thus, discourses are productive; they *do* something that goes beyond simply using signs to represent things and express meaning (Kaarhus 2001: 29). Similar to actor-networks, discursive practices open up and enable some realities while making other options remote or even impossible. Accordingly, Foucault's discourses and ANT are both concerned with the question of representation and *spokespersons*, but, more importantly, the focus is on what is being produced due to a particular kind of material and discursive or epistemic relationality. In *Organizing Modernity*, Law (1994) draws on the Foucauldian notion of discourse when describing the various ways an organization is being patterned and enacted. In his ethnographic study of a scientific laboratory, Law identifies four different logics (administration, enterprise, vision, and vocation) simultaneously at work. These strategies, which he called *"modes of ordering"* (Law 1994), were not just social but materially heterogeneous processes that shaped the organization. Hence, the laboratory was enacted in and through different ordering modes that produced specific forms of organization (Law 2003: 2). While arranging material conditions, knowledge, people, relations, practices, etc., differently, the modes also interacted in what Law describes as a mutual interdependency. In other words, the organization depended on the various modes of ordering for its own continuity and stability. Law's argument is that organization is the outcome of multiple strategies, and there is no single overarching order that holds things together. On the contrary, organizations work and function precisely because of multiple ordering modes and a series of complex relations between them (Law 2003: 2). There is nothing *pure* about modernity, and, therefore, ordering is not pure either but tends to be rather *noncoherent* (Law 2003; Law et al. 2013). The idea of realities as multiple is also extensively developed by Mol, who focuses on how reality is shaped and multiplies in practices. Mol argues that shifting the focus from perspective (seeing from different positions or points of view) to practice allows us to capture the way realities become multiple as the objects that are being manipulated differ along with their practices (Mol 2002: 5). Multiplicity is, therefore, the outcome of several coexisting practices that work and take place simultaneously. This focus on realities as multiple has also generated new ways of approaching policy that question the previous "orthodox and influential notion of policy as a stable object that can be transplanted from one place to another" (Law and Singleton 2014: 381). Instead, from an ANT perspective, "policy is a set of heterogeneous practices done variably in multiple locations" (Law and Singleton 2014: 381).

Later works within ANT have also fruitfully combined Deleuze and Guattari's (1987) assemblage thinking with the conception of actor-networks. The reason is that "both approaches are concerned with why orders emerge in particular ways, how they hold together, somewhat precariously, how they reach across or mould space and how they fall apart" (Müller 2015: 27). There are important differences, however, between the notion of assemblage and actor-network. Müller (2015), for example, emphasizes that ANT considers agency a mediated achievement and, therefore, exclusively the outcome of multiple relations and associations, whereas assemblage thinking assumes that the parts that integrate a new association "can have intrinsic qualities outside associations that impact on and shape the assemblage" (Müller 2015: 31). Besides highlighting that ANT researchers have developed a clearer conception of ANT's relation to politics, Müller also believes the approach offers a more concrete conceptual and methodological framework on how to trace links and associations in empirical studies than assemblage thinking does (Müller 2015: 31). As ANT researchers have ventured into new fields of study, such as public administration, environmental politics, and the economy, the notion of assemblage and actor-networks have been combined in new and interesting ways. Latour's (2005) *Reassembling the Social* is relevant in this context because it constitutes a movement from ANT focusing principally on the production of science and technology towards focusing on society in general. Using the same ANT principles, the aim is to see how society and, more specifically, how "the social" is being produced as a privileged category and a source of explanation. Hence, what Latour proposes is nothing more and nothing less than a new kind of sociology, not of the social but of associations, as society must be reassembled as a new type of collective based on a more symmetrical relationship between humans and nonhumans. This is a political project, which is much broader than what concerns traditional institutional politics.

In the field of economic sociology, Callon has engaged in studies on the relationship between economy, economics, and markets. Callon's point of departure is the concept of *externalities*, which economists use when referring to all those relationships and effects that are not accounted for in a market transaction (Callon 1998b: 16). Externalities are negative when they imply additional costs or investments for actors or entities that find themselves outside the transaction. However, externalities can also be positive, such as when external actors in one way or another benefit from market operations in which they are not involved. Overall, "negative externalities imply social costs that are not taken into account by private decision-makers; positive externalities discourage private investment by socializing the benefits" (Callon 1998a: 248). Within economic theory, the notion of externalities is understood as an example of market failures (the market being inefficient) and, thereby, points to the constructed character of markets (Callon 1998b). The concept of externalities is, therefore, based on *framing*, that is, the separation of relations, actors, goods, objects, commodities, etc., that belong to the transaction from those that are to remain outside. Thus, the very existence of markets and their operability depends on this performative disentanglement. Callon (1998a) explains that he borrows the concept of *frames* from Goffman (1974), and while not employed in economic theory, *framing* is useful to explain a series of investments in "physical and symbolic devices" (Callon 1998a: 252) that are necessary for any kind of performance. Framing, however, can never be complete or totally stable, as it would be too costly to take into account and cover every possible eventuality. It is, therefore, impossible to avoid *leakage*, as the framing process itself constitutes "simultaneously a potential conduit for overflows" (Callon 1998a: 254). Callon sustains that this is not necessarily a negative thing, as overflows participate in circulating and producing new knowledge.

Making a case for a case study of Lofoten and Yasuní-ITT

The ongoing political debate in Norway concerning the possibility of future oil exploitation in Lofoten and an expansion of the petroleum and gas industry in the north presents some elements that are comparable to the Ecuadorian case. The use and the spatial organization of the territory is a major part of the debate in both countries. Many stakeholders in Norway believe that coexistence is impossible between the fisheries, tourism, and the oil industry in the same territory. Similarly, a conflict over land use and territorial rights is also part of the Ecuadorian debate. In the Amazon, the coexistence between indigenous communities, agriculture, logging, and the oil companies has been anything but harmonious. In short, both cases are part of larger national debates related to the expansion of the oil frontier and, consequently, possible social, environmental, and climate implications.

Nevertheless, my decision to select Lofoten and Yasuní-ITT was based on their usefulness as contrasting cases. Furthermore, including a case from a developed country like Norway and another case from a developing country like Ecuador could produce important insights into the challenges of non-extraction policies and place them in a more general global context. What makes Yasuní-ITT and Lofoten contrasting cases is that, among other things, the Norwegian and the Ecuadorian oil histories (and, therefore, also narratives) have produced substantially different outcomes. The two countries are usually placed at different ends of the spectrum in the resource curse literature. The *resource curse hypothesis* sustains that the combination of economic, political, and environmental pathologies that some countries experience can have a cumulatively negative effect on the national economy and, as a consequence, these societies are unable to benefit from the revenues generated by their energy exports (Bradshaw 2014: 157). Norway, together with countries like Canada, is referred to as a country that has managed to avoid *the resource curse* and has used its oil and gas resources to generate economic development (Stevens 2003; Davis and Tilton 2005), while Ecuador's petroleum experience for several decades has had the opposite effect on its national economy, exhibiting characteristics typically associated with Dutch disease.[1]

Selecting two cases that apparently had little in common (at least, at first sight), other than belonging to oil-exporting countries, offered me the opportunity to search for similarities that did not necessarily stand out as obvious. However, despite being different, the cases evidently share a broader context that includes, among other things, a common global climate regime and the international oil market. This is important since the dramatic drop in oil prices that started in June 2014 and the signing of a new climate agreement in Paris at the end of 2015 are events that influenced and informed both cases and, therefore, became key variables I had not foreseen when I first wrote my research proposal. However, sampling or selecting cases is also a matter of *making* some relations or links commensurable. As Strathern has pointed out, "comparability is not intrinsic to anything" (2004: 53). Developing a point made by Howe (1987), Strathern emphasizes that comparison is not about finding things that are similar or different in themselves to compare them; instead, it is about acknowledging that the process of selecting and comparing is what produces relations of similarity and difference (Howe 1987 in Strathern 2004: 53).

The complexity of the cases made me realize that a more traditional approach towards comparative case studies would impose certain limitations on the definition of the cases and, thereby, on the analytic possibilities, as most case study literature emphasizes "bounding" as a key requirement for comparison (see Creswell 2013; Yin 2014). Since Lofoten and Yasuní are ongoing and evolving political processes, working with fixed and clearly delimited cases seemed problematic as these controversies are distributed and occur at many different locations simultaneously. I therefore ended up using an approach, which is more in line with what Bartlett and Vavrus refer to as a comparative case study heuristic that draws on both multi-sited ethnographies developed within contemporary anthropology (Bartlett and Vavrus 2017) and on process-oriented qualitative studies (Maxwell 2013 in Bartlett and Vavrus 2017). Accordingly, a *processual* comparative case study allowed me to trace and analyze relationships of similarity and difference between sites, which implied focusing on what Bartlett and Vavrus (2017) call the *horizontal axis*, while simultaneously paying attention to how policies, processes, and events developed across scales, which makes up the *vertical axis*. According to Bartlett and Vavrus, "the vertical axis reminds us to follow the phenomenon itself, be it a practice or a policy, as it enlists and engages actors whom one might otherwise assume operate in bounded spaces" (2017: 74). Finally, a processual comparative case study approach gave me the opportunity to work with a different conceptualization of context, not as a fixed place or location, but "as something spatial and relational" (Bartlett and Vavrus 2017: 47), which operates across scales and is woven into and, thereby, becomes part of the cases in various and shifting ways.

The cases

The Lofoten Islands, located in the Norwegian Sea above the Arctic Circle, are the center of important cod fisheries that take place every year between January and April. The fisheries and the production of stockfish have historically played a decisive role in the settlement and the local communities along the coast of northern Norway. The fisheries have generated important economic growth and development in the region. The waters around Lofoten and Vesterålen are the main spawning areas for cod. According to the Norwegian Institute of Marine Research (2019), "the Northeast Arctic cod stock is the largest cod stock in the world", and an acute oil spill from a major incident would certainly have negative consequences for the marine environment and the fragile ecosystems. Past research suggested that 6% of all cod and herring larvae would disappear with an acute oil spill. The Institute of Marine Research later adjusted its calculations, indicating that as much as 100% of a larvae year-class could disappear in the case of a major spill. The oil industry, however, is rather skeptical of the calculations presented by the institute since they consider them based on a worst-case scenario and therefore not very realistic (Haraldsen 2009).

Disagreements regarding the levels of environmental risk tied to oil activity in the waters of Lofoten, Vesterålen, and Senja (LoVeSe) have also generated considerable political debate. As for today, two of the major Norwegian political parties[2] agree on opening up the area to the petroleum industry. However, several minority parties, together with environmental organizations, are strongly against oil drilling in LoVeSe not only because of the fisheries and the importance of the marine ecosystems but also because of climate change. The argument is that opening up new areas for oil exploitation will increase Norwegian CO_2 emissions and prolong the country's oil dependency for several decades instead of starting to phase-out the industry. They argue that Lofoten is symbolically important to show that Norway's true intention is to contribute to an energy shift based on renewables (Hersoug and Arbo 2010: 317). Another element in the extraction or non-extraction debate has been the possibility of getting Lofoten on UNESCO's World Heritage List as a mixed site due to its natural and cultural values. This process has been stopped, as some actors feared that a UNESCO status could block the possibility of future oil activity in the area.

Yasuní-ITT is located in the western part of the Amazon Basin, approximately 250 kilometers from Ecuador's capital, Quito. Yasuní National Park, declared a World Biosphere Reserve by UNESCO in 1989, is one of the most biodiverse places on earth, as well as the territory of two uncontacted, indigenous groups, the Taromenane and the Tagaeri, who live in voluntary isolation in the rainforest. During the United Nations General Assembly in 2007, Ecuador's President Rafael Correa launched a new idea: The country would "leave the oil in the soil" (Martin 2011; Espinosa 2013) in exchange for financial compensation from the international community. In his speech, Correa argued that developing countries with important biological and cultural diversity should be able to leave their petroleum resources underground and receive economic support from industrialized countries. By keeping 846 million barrels of oil in the ground indefinitely, Ecuador would prevent 407 million metric tons of CO_2 from being released.[3] This project, known as the Yasuní-ITT Initiative, became a reality in 2010 with help from the United Nations Development Programme (UNDP). The goal was to receive 3.6 billion dollars over a 13-year period, which Ecuador would use in different strategic areas, such as the development of renewable energy sources, reforestation, social development programs, research, and technology. After years of campaigning, the initiative failed and, in August 2013, the Ecuadorian government announced that it had decided to end the project due to the lack of contributions and start drilling for oil in the Yasuní-ITT field.

Methodology and data collection strategies

One of the principal strengths of a case study is that it allows the researcher to use many different sources of data and data collection techniques. A single source of information is usually not enough to address the complexities of this kind of inquiry (Creswell 2013; Yin 2014). Accordingly, several sources of data were combined during the data collection process for this project. Before I started fieldwork at the beginning of 2015, I spent several months collecting and reading news articles, documents, reports, and web pages related to the cases to obtain necessary background information. Since the study focuses on current processes, local stakeholders and national actors are constantly positioning both themselves and the perspectives and knowledge systems they consider relevant. Personally, I found this situation challenging. In fact, since a variety of circumstances and political situations changed during the time I was carrying out the research, there was always new information that, in one way or another, was highly relevant for the project. Following these events, however, was vital for several reasons. First, media coverage, reports, and documents helped me keep track of the political processes and, thereby, acquire important background knowledge of how values, facts, arguments, and discourses were produced and circulated among different stakeholders. Furthermore, it also served as an introduction to important themes and their chronological development in both cases. In short, what happened when, and who and what was involved. Second, I also got an overview of relevant actors who participated in the national debate on petroleum and climate issues. This was very useful further down the road when I was making a list of possible informants. Finally, the insights I gained by closely following the debate in the media and through different types of documents directly informed the interview process since part of this information was included in the interview guides. On several occasions, I would specifically refer to or ask about a statement put forward by a politician, an environmentalist, or a government official in order to obtain the informants' perspective on the issue in question. Keeping track of what was happening through media coverage and documents was, therefore, important throughout the whole research process, in other words, before, during, and after fieldwork. Besides fulfilling the role as key sources of information, documents and reports also progressively acquired a more active role during the research process, namely that of "agents in networks of action" (Prior 2008: 822). I was specifically interested in how documents shaped and directly participated in the socioenvironmental conflicts in question. Particular attention was, therefore, given to documents that helped move the processes in a specific direction, worked to constitute realities, or linked actors and sites.

In combination with documentary research, the data material is the result of 50 qualitative interviews carried out with 25 informants in Ecuador and 25 informants in Norway. These interviews were all recorded and later transcribed. According to Berg and Lune, interviews are useful when the researcher is interested in finding out how participants perceive situations and how they "attach certain meanings to phenomena or events" (2012: 115). Moreover, I used an adaptation of what is often referred to as a *semi-structured* format to conduct the interviews. The advantage is that while using a prepared interview guide with planned questions and themes, the researcher also has the opportunity to use probes and follow up on important information that was not necessarily included in the interview guide originally.

Thus, between one interview and another, I constantly modified and updated the interview guide since some questions did not seem relevant or appropriate for certain informants or actually turned out to not be relevant at all, while other questions emerged during the interview process. Some questions and themes were the same for all interviews (just with slightly different wording), while others were deleted as new promising themes were introduced. Since the group of informants included a variety of different actors with very different roles and involvement with the cases, I constantly had to prepare new interview guides with this in mind in order to gain access to specific information that a particular informant was in a position to provide.

The process of interviewing was, therefore, a continuous process. While each interview was a unique event with its own particularities, it was also part of a broader interviewing process. It is fruitful, therefore, to consider the interviews as interconnected in an explorative, dynamic, and cumulative process (Kaarhus 1999: 57). Hence, the interview as data material cannot be looked at in isolation but as part of a series of informational events. This perspective is precisely what allows the researcher to look for common patterns or themes across the interviews or across several cases in a multiple or comparative case study.

Another important aspect that is worth mentioning is the active role the informants played in shaping the research process. Some of the informants who participated in the inquiry generously gave me, after finishing the interview, a variety of documents like books and booklets, flyers, articles, or they would later send me links with information. On other occasions, they would specifically mention publications like research reports, videos on YouTube, and other documents and where I could find them. Evidently, they did this because they thought it was relevant for the project and that I should read the material they gave me. This clearly shows that the informants, besides giving information during the interviews, often actively provided additional sources of information and, thereby, extended their participation beyond the interview situation itself. Evidently, informants also shaped the interviews and interview situation in other ways. On several occasions, I had to abandon the themes that I had prepared in my interview guide because I realized that what the informant was telling me was so important that the best thing I could do was to follow up with exploratory questions and probes. Sometimes, if I had the opportunity, I would return to the guide later on in the interview if I still felt that the informant had not covered some of the relevant topics. The sequence and structuring of the interview were, therefore, very much a product of the interview situation and the interaction between the interviewee and myself. For this reason, I do not agree with the conception of the research interview as an a priori asymmetrical relationship between an interviewer and an interviewee (Kvale and Brinkmann 2015: 51–52), in which the former has the power to completely define the interview situation. For instance, the setting or staging of the interviews was also mostly up to the informants since they chose where and when the interviews would take place. For this reason, the setting varies enormously from very formal contexts, like offices or conference rooms in the Norwegian parliament or the Ministry of Climate and Environment, to informal settings like someone's home or a small, noisy street corner cafeteria in Quito. In fact, sometimes, informants would reschedule the appointment or change the place for the interview because they thought it was better or something else came up that they had to attend. On other occasions, they would ask me where I thought it would be okay to meet, suggesting, in other words, that I should select "a proper" setting for the interview. The setting is, therefore, by no means defined a priori only by the researcher but appears as a negotiable dimension both before and during the interview (Kaarhus 1999: 40). These situations clearly illustrate that informants are not just passive entities who provide answers that the interviewer evokes by using questions prepared beforehand. On the contrary, the previous examples demonstrate that they engage in a role as co-producers of the interview in a variety of ways.

Fieldwork

Fieldwork, which constitutes the core data material for this inquiry, was conducted in two stages: First in Ecuador between January and June 2015 and later in Norway from September 2015 to January 2016. I interviewed local and national actors in both cases. This was important, as I was interested in obtaining diverse perspectives on different levels. The reason for conducting fieldwork in Ecuador first, however, was mainly pragmatic because I thought it would be easier to start back home since I had more knowledge and, therefore, a better understanding of the Ecuadorian case than the Norwegian one. Additionally, it would give me more time to upgrade my knowledge on the oil debate in Norway before starting the interviewing process.

I applied a combination of purposive and snowball sampling (Blaikie 2010) to recruit informants; this means that some participants were selected based on my knowledge about their involvement in the cases, while others were contacted because other participants considered they were relevant for the inquiry and, therefore, should be included. While recruiting informants in Ecuador, one important criterion was to incorporate informants who had been involved in the Yasuní Initiative in an earlier stage. Other informants were selected because they actively participated in the current debate. The reason for this strategy was that it allowed me to incorporate informants with different points of view from different phases or stages of the political process of the Yasuní-ITT Initiative.

I normally used the internet to contact and inform possible informants about the inquiry if their e-mail address was available on their organization's website. If this was not the case, I would contact them by phone and ask for an appointment after explaining the project's background and objectives. Since some of the informants did not belong to formal organizations or organizations with updated websites, e-mail addresses were not always available. This was never a problem in Norway, but in Ecuador, some participants were not easy to locate. Therefore, I had to make use of my personal network of family, friends, and former colleagues who helped me locate people or, as it often turned out, locate other people who could help me contact the person in question. Informants whom I had previously interviewed were also very helpful, providing e-mail addresses and phone numbers of people I wanted to contact.

Another difficulty I encountered while recruiting participants in Ecuador was that some government officials and people with close ties to the government never replied to my e-mails or directly declined to participate in the inquiry. After several attempts, I realized that gaining access was not possible and, therefore, started to look for other informants who could voice the government's perspective in some way. I believe the reason for this situation was the growing tensions between the government and certain civil society groups that originated when environmentalists and indigenous groups accused the National Election Council (*Consejo Nacional Electoral*) of fraud when thousands of signatures were rejected in May 2014, blocking the possibility of a national referendum on oil extraction in Yasuní (The Guardian 2014). Distrust on behalf of the government also grew when members of the German parliament tried to enter the national park in December 2014 without an official permit from the Ecuadorian authorities (BBC Mundo 2014). Against this background, I believe some people felt that giving an interview was complicated and, therefore, decided not to participate. What is more, on two occasions, people directly told me that the problem with my research project was that it involved a politically sensitive issue.

Furthermore, in Ecuador, where I had followed the case and its development for several years, I could ask more specific and detailed questions. In Norway, however, I did not have the same familiarity or prior knowledge regarding the petroleum debate. The solution was to employ a set of more general questions, at least in the beginning. After a while, I was able to use questions that were a lot more specific and formulate follow-up questions in a more detailed manner. Another strategy that turned out to be quite useful was that I decided to provide an explanation about my background before I started each interview. The reason for this was that I feared that I would lose important shared or common knowledge about the case if the informants erroneously believed I already knew things that "everybody" knows and excluded this type of information from their accounts or answers. Therefore, I briefly explained to the interviewees that I grew up in Ecuador, where my family moved when I was a little girl, and that this was the reason for me not necessarily knowing what everybody knows about the

Norwegian oil debate regarding Lofoten. As I see it, this disclosure about myself made informants incorporate more details in their accounts as they became aware of certain background information that I was perhaps lacking. Some interviewees also made drawings of the territory to show me how specific characteristics of the topography and ocean currents in this area made the region highly vulnerable in the case of a major oil spill or a blowout. An informant also drew the movement of different currents in the area to explain why he believed Statoil's (now Equinor) data on this issue were inaccurate since he knew the area from his work as a fisherman for more than 20 years. In other words, he depicted his experience-based knowledge as opposed to the national oil company's use of science-based knowledge. How these two knowledge systems are enacted and used in the petroleum debate will be analyzed in Chapter 3. Moreover, another informant drew a timeline to explain the chronology of important political events during the struggle to keep the territory outside LoVeSe closed to petroleum activity. The informants' drawings had the ability to connect and visualize events and environmental risk. By drawing locations, sensitive areas, currents, and *Eggakanten* (the edge of the continental shelf), etc., invisible and intangible risks became connected and specified on paper. The drawing activity linked experience, events, and context, as the informants were able to accompany some of their explanations with visual elements and, thereby, enhance my understanding. In other words, the drawings became important vehicles for spatial and temporal positioning of environmental and political challenges according to the informants' perspectives. These are examples that point to the importance of material conditions during the interview situation.

In Ecuador, besides conducting in-depth interviews, a short field trip to Tiputini Biodiversity Station, owned by the University of San Francisco in Quito, in collaboration with Boston University, was also important for the project. The research station, located within the Yasuní Biosphere Reserve in the Eastern Ecuadorian Amazon, is internationally recognized as a unique place to study biodiversity and tropical ecosystems due to its isolation and pristine rainforest. [4] Before my trip, I had agreed not to conduct any formal interviews during the visit at the station since all research projects require an official permit from the Ministry of the Environment in Ecuador, something I considered out of the scope of this inquiry since I was not going to carry out systematic observations or data collection. Alternatively, my purpose was to better understand the importance of the conservation of the rainforest and its ecosystems. The trip provided me with a unique opportunity to observe a variety of species in their natural habitat, feel the extreme humidity while walking through the forest with other students and the guides who worked at the station, and, last but not least, see what healthy rainforest looks like. In short, I had a very basic understanding of the concept of biodiversity before arriving at the station in Tiputini; however, the visit transformed my understanding into something experiential, as there was life everywhere. My approach was rather simple. I felt that in order to write about the challenges related to oil extraction in highly sensitive environmental areas, I had to visit these territories, even if only for a short time, to gain an experiential intake of the complexities of the Amazonian rainforest and the Lofoten Islands.

In the case of Lofoten, I had the opportunity to visit the islands several times during data collection; however, my first visit was in September 2014 during the yearly Science Week when Norwegian universities and research institutes present their projects and activities to the public. I participated in a series of seminars with a group of professors and researchers from the Faculty of Social Sciences at Nord University. In Reine, a fishing community that belongs to the Moskenes municipality, located above the Arctic Circle, we slept in traditional fishing cabins and tried the world-famous bacalao. At night, the Northern lights or *aurora borealis* lit up the dark sky over the harbor, and I remember the experience as truly magical. Being able to visit and see the places and the inherent characteristics in which the controversies regarding oil extraction in these regions have originated has been crucial, as I do not believe it is possible to carry out case study inquiry without some sort of contact or first-hand experience with the cases' geographic locations.

Chapter outline

Chapter 1 introduces the reader to the book's topic, the questions it addresses, its arguments, and analytical framework. The purpose is to position the inquiry within existing international climate debates and discussions. The chapter also explains the rationale behind selecting Lofoten and Yasuní and their usefulness as contrasting cases. Finally, it discusses the inquiry's methodology and the data material the analysis rests on. Chapter 2 analyzes how global uncertainties arising from climate science inform the Norwegian petroleum debate and how controversies and processes related to oil extraction in environmentally sensitive areas travel between conflicting logics and arguments. Drawing on Law (1994), I argue that safety and insecurity can be understood as two different modes of ordering regarding Norway's oil-driven economy. The analysis suggests that behind the discourse and storylines about the country's "petroleum adventure", there are both contradictions and ambivalence. Focus is placed on how Norway actively uses different framings to reconcile its role as an oil and gas producer with its ambition of being a forerunner in international climate negotiations. By enacting climate change as an economic issue, embedded in market assumptions, Norway has managed to politically separate petroleum production from climate change, despite scientific research linking them together. The last section examines the main factors that are delaying Norway's "green shift" towards a more diversified and less oil-dependent economy. Chapter 3 addresses the ongoing sociopolitical process and environmental controversy attached to the possibility of opening up the Lofoten area to oil drilling and, thereby, expanding the oil frontier towards territories that are classified as "valuable" and "vulnerable" in several research documents and the integrated management plan. The analysis suggests that these categories have become important means of negotiation and resistance among different stakeholders. Overall, people do not question the authority of the existing knowledge base. As such, scientific knowledge becomes an important *ally* in the petroleum debate. What is negotiated, however, is the political implications and practical meanings of these categories, that is, what kinds of *matters of concern* (Latour 2005) emerge and are enabled due to these categorizations. As part of this analysis, the chapter looks into how tensions between science and politics and oil and environment are enacted and translated. The last section is dedicated to the analysis of Lofoten as a heterogeneous territory and how this situation produces conflicting and overlapping network translations.

Chapter 4 starts out by examining Ecuador's oil history from a "curse" perspective, which constitutes a frame for ordering the multiplicity of oil. Emphasis is placed, therefore, on how the Amazon crude produces global connectivity and local fragmentation simultaneously. Here, I identify and discuss three modes of ordering related to oil extraction in Amazonia: Developmentalism, environmental destruction, and violation of rights. I then examine some key socioenvironmental conflicts, focusing on territorial disputes as the outcome of spatial dynamics rooted in the underground petroleum reservoirs. The chapter's last section describes the inscription of Nature(s) in the Ecuadorian Constitution. Chapter 5 addresses the process of the Yasuní-ITT Initiative as an evolving design that assembled and enrolled a series of elements and actors in an attempt to "go global". Callon's (1998a) pairing concepts of framing and overflowing are used to examine to what extent the Yasuní Initiative overflowed the UNFCCC and the Kyoto Protocol as it attempted to open the "black box" of climate change mitigation. Emphasis is placed on tracing some of the policy documents and the different parallel processes that were assembled in the Yasuní-ITT Initiative. This chapter also analyzes how the struggle to keep the oil in the ground was translated as a matter of defending the rights of humans and nonhumans.

In Chapter 6, after a brief introduction on oil and climate as intersecting epistemic objects, I draw the cases together by comparing and contrasting key elements and themes from the previous chapters. Lofoten and Yasuní, as interfering networks, constitute translocal political processes, which implies tracing links and connections across local, national, and global levels. By comparing the management plans for the Barents Sea–Lofoten area and Yasuní National Park, the chapter identifies several similarities and differences in the way territorial disputes are domesticated through zoning and how the underground petroleum reservoirs become a particularly important drive in this process. Hence, the chapter provides a cross-case analysis of how various encounters between oil and climate influence aspects such as territorializations and spatial ordering, the framing of climate change, and the production of both natures and socioenvironmental risks. In Chapter 7, I revisit the empirical chapters and discuss some important analytic choices; I then sum up the main findings in light of the initial questions presented in the first chapter and discuss possible contributions and limitations of the inquiry. The chapter concludes with a final comment regarding the analytic potential of the processual comparative case study methodology used in the study.

Notes

1　The term "Dutch disease" refers to a situation where the non-resource sector of the economy has become uncompetitive on the international market. The appreciation of the national currency stimulates the import of goods and, consequently, the domestic non-resource sector declines (Bradshaw 2014: 173).
2　While the Conservative Party and the Progress Party wish to open LoVeSe for petroleum activity, the Labour Party changed its position in 2019.
3　UNDP Multi-Partner Trust Fund Office: Yasuní-ITT Fact sheet: http://mptf.undp .org/yasuni.
4　Boston University, Center for Ecology & Conservation Biology. http://www.bu.edu/ cecb/tiputini/.

References

Asdal, K., Brenna, B., & Moser, I. (2007). The politics of interventions: A history of STS. In: K. Asdal, B. Brenna, & I. Moser (eds), *Technoscience: The Politics of Interventions* (pp. 7–57). Oslo: Unipub.

Bartlett, L., & Vavrus, F. (2017). *Rethinking Case Study Research: A Comparative Approach*. New York: Routledge.

BBC Mundo (2014, December 14). La visita alemana que no es bienvenida por el gobierno de Ecuador. Retrieved from https://www.bbc.com/mundo/noticias/2014/12/14 1209_ecuador_alemania_yasuni_cch.

Berg, B. L., & Lune, H. (2012). *Qualitative Research Methods for the Social Sciences*. Boston: Pearson.

Blaikie, N. (2010). *Designing Social Research*. Cambridge: Polity Press.

Blok, A., & Jensen, T. E. (2011). *Bruno Latour: Hybrid Thoughts in a Hybrid World*. London: Routledge.

Bloor, D. (1976). *Knowledge and Social Imagery*. London: University of Chicago Press.

Boston University (n.d.). Tiputini Biodiversity Station. Retrieved from Center for Ecology & Conservation Biology https://www.bu.edu/cecb/tiputini/.

Bradshaw, M. (2014). *Global Energy Dilemmas*. Cambridge: Polity Press.

Callon, M. (1986). Some elements of a sociology of translation: Domestication of the scallops and the fishermen of St. Brieuc Bay. In: J. Law (ed), *Power, Action and Belief: A New Sociology of Knowledge?* (pp. 196–233). London: Routledge & Kegan Paul.

Callon, M. (1995). Four models for the dynamics of science. In: S. Jasanoff, G. E. Markle, J. C. Petersen, & T. Pinch (eds), *Handbook of Science and Technology Studies* (pp. 29–63). Thousand Oaks: Sage.

Callon, M. (1998a). An essay on framing and overflowing: Economic externalities revisited by sociology. In: M. Callon (ed), *The Laws of the Markets* (pp. 244–269). Oxford: Blackwell Publishers.

Callon, M. (1998b). Introduction: The embeddedness of economic markets in economics. In: M. Callon (ed), *The Laws of the Markets* (pp. 1–57). Oxford: Blackwell Publishers.

Chevalier, J.-M. (2009a). Introduction. In: J.-M. Chevalier (ed), *The New Energy Crisis: Climate, Economics and Geopolitics* (pp. 1–5). Basingstoke: Palgrave Macmillan.

Chevalier, J.-M. (2009b). The new energy crisis. In: J.-M. Chevalier (ed), *The New Energy Crisis: Climate, Economics and Geopolitics* (pp. 6–59). Basingstoke: Palgrave Macmillan.

Creswell, J. W. (2013). *Qualitative Inquiry & Research Design: Choosing among Five Approaches*. Los Angeles. Sage.

Davis, G. A., & Tilton, J. E. (2005). The resource curse. *Natural Resources Forum*, 29(3), 233–242.

Deleuze, G., & Guattari, F. (1987). *A Thousand Plateaus: Capitalism and Schizophrenia.* Minneapolis, MN: University of Minnesota Press.

Espinosa, C. (2013). The riddle of leaving the oil in the soil: Ecuador's Yasuní-ITT project from a discourse perspective. *Forest Policy and Economics*, 36, 27–36.

European Commission (2007, October 24). Towards a post-carbon society: European research on economic incentives and social behaviour. Retrieved from https://espas.secure.europarl.europa.eu/orbis/sites/default/files/generated/document/en/Post%20Carbon%20Society.pdf.

Giddens, A., & Sutton, P. W. (2013). *Sociology.* Cambridge: Polity Press.

Goffman, E. (1974). *Frame Analysis: An Essay on the Organization of Experience.* Boston: Northeastern University Press.

Haraldsen, I. (2009, August 27). *Faglig strid om olje i nord.* Retrieved from http://forskning.no/fisk-havforskning-olje-og-gass/2009/08/faglig-strid-om-olje-i-nord.

Hersoug, B., & Arbo, P. (2010). Petroleumsvirksomhet i nord: Redning eller trussel? In: P. Arbo, & B. Hersoug (eds), *Oljevirksomhetens Inntog i Nord: Næringsutvikling, politikk og samfunn* (pp. 305–330). Oslo: Gyldendal Akademisk.

Hess, D. J. (1997). *Science Studies: An Advanced Introduction.* New York: New York University Press.

Howe, L. (1987). Caste in Bali and India: Levels of comparison. In: L. Holy (ed), *Comparative Anthropology.* Oxford: Blackwell.

International Energy Agency (2012). World energy outlook 2012: Executive summary. Retrieved from http://www.iea.org/publications/freepublications/publication/English.pdf.

Institute of Marine Research (2019, July 19). Topic: Cod – Northeast Arctic. Retrieved from https://www.hi.no/en/hi/temasider/species/cod--northeast-arctic.

Kaarhus, R. (1999). Intervjuer i samfunnsvitenskapene: Bidrag til en videre metodologisk Diskurs. *Tidsskrift for Samfunnsforskning*, 1, 33–62.

Kaarhus, R. (2001). En Foucault-inspirert diskursanalyse. *Sosiologi i Dag*, 31(4), 25–46.

Kristoffersen, B., & Dale, B. (2014). Post petroleum security in Lofoten: How identity matters. *Arctic Review on Law and Politics*, 5(2), 201–226.

Kvale, S., & Brinkmann, S. (2015). *Det Kvalitative Forskningsintervju.* Oslo: Gyldendal Akademisk.

Latour, B. (1993). *We Have Never Been Modern.* Cambridge, MA: Harvard University Press.

Latour, B. (2005). *Reassembling the Social: An Introduction to Actor-Network Theory.* New York: Oxford University Press.

Law, J. (1987). Technology and heterogeneous engineering: The case of the Portuguese expansion. In: W. E. Bijker, T. P. Hughes, & T. Pinch (eds), *The Social Construction of Technical Systems: New Directions in the Sociology and History of Technology* (pp. 111–134). Cambridge, MA: MIT Press.

Law, J. (1994). *Organizing Modernity.* Oxford: Blackwell.

Law, J. (2003). Ordering and obduracy. Retrieved from Centre for Science Studies. Lancaster University: https://www.lancaster.ac.uk/fass/resources/sociology-online-papers/papers/law-ordering-and-obduracy.pdf.

Law, J. (2007). Actor network theory and semiotics. Retrieved from http://heterogeneities.net/publications/Law2007ANTandMaterialSemiotics.pdf.

Law, J., & Singleton, V. (2014). ANT, multiplicity and policy. *Critical Policy Studies*, 8(4), 379–396.

Law, J., Afdal, G., Asdal, K., Lin, W.-y., Moser, I., & Singleton, V. (2013). Modes of syncretism: Notes on noncoherence. *Common Knowledge*, 20(1), 172–192.

Lidskog, R., & Sundqvist, G. (2013). *Miljøsosiologi*. Oslo: Gyldendal Akademisk.

Martin, P. L. (2011). Global governance from the Amazon: Leaving the oil underground in Yasuní National Park, Ecuador. *Global Environmental Politics*, 11(4), 22–42.

McGlade, C., & Ekins, P. (2014). Un-burnable oil: An examination of oil resource utilisation in a decarbonised energy system. *Energy Policy*, 64, 102–112.

Mol, A. (2002). *The Body Multiple: Ontology in Medical Practice*. Durham: Duke University Press.

Müller, M. (2015). Assemblages and actor-networks: Rethinking socio-material power, politics and space. *Geography Compass*, 9(1), 27–41.

Newton, T. (2007). *Nature and Sociology*. New York: Routledge.

Princen, T., Manno, J. P., & Martin, P. L. (2015). The problem. In: T. Princen, J. P. Manno, & P. L. Martin (eds), *Ending the Fossil Fuel Era* (pp. 3–36). Cambridge, MA: MIT Press.

Prior, L. (2008). Repositioning documents in social research. *Sociology*, 42(5), 821–836.

Shapin, S., & Schaffer, S. (1985). *Leviathan and the Air-Pump: Hobbes, Boyle, and the Experimental Life*. Princeton, NJ: Princeton University Press.

Stern, N. (2007). *The Stern Review: The Economics of Climate Change*. Cambridge: Cambridge University Press.

Stevens, P. (2003). Resource Impact: Curse or blessing? A literature survey. *Journal of Energy Literature*, 9(1), 3–42.

Strathern, M. (2004). *Partial Connections*. Walnut Creek: Rowman & Littlefield Publishers.

The Guardian (2014, May 8). Ecuador rejects petition to stop oil drilling in Yasuní National Park. Retrieved from https://www.theguardian.com/environment/2014/may/08/ecuador-rejects-petition-oil-drilling-yasuni.

UNDP Multi-Partner Trust Fund Office (n.d.). *Yasuní-ITT Fact Sheet*. Retrieved from http://mptf.undp.org/yasuni.

Yin, R. K. (2014). *Case Study Research: Design and Methods*. Thousand Oaks: Sage.

2 Norwegian oil as a provider of safety and/or (future) insecurity

The relationship between climate change and fossil fuels is characterized by ambiguity and scientific abstractions to which citizens do not necessarily find it easy to relate. After a brief presentation on challenges related to our current energy system, I will analyze how these global uncertainties inform the Norwegian petroleum debate and how controversies and processes related to oil extraction in highly sensitive areas travel between conflicting logics and arguments. Moreover, I address the interplay between politics, knowledge, and technology to show how different actors and entities enact oil as safety and/or insecurity. Finally, I analyze the factors that delay Norway's *green shift* towards a more diversified and less oil-dependent economy.

In his book *Global Energy Dilemmas*, Bradshaw (2014), a professor of global energy, states that it has become

> evident that over the last two centuries there has been a dramatic change in the relationship between energy and the development of human society. The harnessing of a geological storehouse of the sun's energy via the evolution of the fossil-fuel system has enabled unimaginable advances in the quality of living for many on the planet. Unfortunately, this fossil-fueled economic miracle has come at considerable ecological cost.
>
> (2014: 12)

This statement has at least two simple but, nonetheless, important implications from an actor-network theory (ANT) perspective. First, environmental degradation is an unintended effect of particular associations between humans and the energy resources we use. The extraction and burning of fossil fuels is part of a system that has managed to become more or less stable in an extended and expanding network of pipelines, geopolitics, petroleum geologists, oil markets, drilling technology, importing and exporting countries, carbon-intensive industry, the Organization of the Petroleum Exporting Countries (OPEC), gasoline cars, oil reserves, refineries, oil companies, the American Petroleum Institute (API) gravity, etc. What is new, however, is that climate change is "forcing" its way into that very same network through research reports, carbon budgets, and international organizations like the Intergovernmental Panel on Climate Change (IPCC) and the International Energy Agency (IEA). This means that climate research is making fossil fuels problematic by pointing out the unsustainability of emissions over a certain threshold. In other words, oil (as well as coal and gas) is becoming increasingly uncertain and has, therefore, more or less turned into a *matter of concern* (Latour 2005).

Second, it is becoming increasingly difficult to talk about the development of human society as the development of a separate domain. As several actor-network theorists have argued, what we usually refer to as "society" is more of a composition with a rather heterogeneous makeup than an integrated "whole". Consequently, the term "social" within this approach bears no reference to a particular substance or essence of reality but constitutes "a way of tying together heterogeneous bundles, of *translating* some type of entities into another" (Latour 2000: 113). Accordingly, phenomena such as climate change or the Anthropocene are not only about how humans are drastically changing the planet and the atmosphere but also, and more precisely, about possible long-term effects generated by multiple ways of interacting between humans and nature. The linkages and assembling processes are often enabled through and by technology. As an example of how science and technology, humans, and nature are engaged in co-producing each other, I include the following answer from a senior adviser in a Norwegian government agency when asked about the role of technology when dealing with environmental issues:

> I think it is a very important criterion because technology is an important premise regarding the environment, cleaning technology, the way we use engineers and technology to take care of the ecosystems, we have natural scientists and biologists who are able to understand connections in nature and why and how they can be important and how we can interact with them. That is the way we humans are created. I do not think that we will ever say now we will let nature arrange (organize) itself. In a certain way we are beyond that point.

This excerpt points to an understanding of nature as partially the outcome of processes in which humans are profoundly involved. Moreover, by organizing and rearranging nature in particular ways through resource extraction, transformation, and consumption, there are often certain externalities or side effects that go far beyond the expected socioeconomic outcomes. This is very much the case if we examine the global energy system. While the extraction and commercialization of hydrocarbons provide exporting countries with important revenues that constitute a considerable share of gross domestic product (GDP) and total exports in many cases, severe environmental deterioration and climate change are also part of the equation. However, the world's energy demand is by no means declining; on the contrary, the IEA expects the energy consumption worldwide to grow by one-third towards 2040 (2015b: 1). Whereas the demand growth in Organization for Economic Co-operation and Development (OECD) countries seems to have stabilized after peaking in 2007 (IEA 2015b: 1), the current increase in demand is primarily located in Asia, especially in emerging economies like China and India. Access to secure and affordable energy supplies is an important factor for achieving economic growth and development. As for today, this still means relying heavily on fossil fuels that comprise more than 80% of global energy production (IEA 2020). This situation, however, highlights the urgent "need to transit towards a low carbon future" (Goldthau 2011: 213). Although governments and policymakers are taking positive steps and increasingly incorporating renewables into their countries' energy mix, this is not happening fast enough. According to the IEA's projections, "energy policies, as formulated today, lead to a slower increase in energy-related [carbon dioxide] CO_2 emissions, but not the full de-coupling from economic growth and the absolute decline in emissions necessary to meet the 2°C target" (IEA 2015b: 7).

There is also insecurity tied to former ideas about future conflicts due to shrinking global petroleum reserves and a possible race to secure new fields. In recent years, the development of innovative technological solutions and the increased use of hydraulic fracking (primarily in the US and Canada) has radically changed not only the former scarcity view but also the energy market and, thereby, the geopolitics of oil. On the one hand, there is significant growth in shale gas and oil production, which has added substantially to global energy reserves. On the other hand, future carbon restrictions due to climate commitments in a 2°C scenario will most certainly cause oil demand to decline, resulting in global oil abundance (Verbruggen and Van de Graaf 2015: 21). Excess oil could create a completely different situation than the one usually portrayed: One possibility is that oil-exporting countries will try to extract and sell as much oil as possible before the reserves lose their value. Another strategy could be to try to keep rival producers' oil in the ground while competing for the remaining quota in a carbon-constrained market (Verbruggen and Van de Graaf 2015: 22). In other words, we are witnessing a new and different energy crisis: "The new crisis comes from the recent intrusion of climate change issues into energy economics and geopolitics" (Chevalier 2009: 1). Additionally, the energy and climate challenges the world faces today expose the interconnectedness of a series of uncertainties: Local becomes global and vice versa. An example of this situation is the way externalities from production and the use of energy resources "no longer remain local or regional but become global instead" (Goldthau 2011: 214). Scales are, therefore, by no means fixed but transit and transform as new elements, calculations, and simulated trajectories inform the global climate regime. While climate change is recognized and proven as real by the scientific community, "no one knows exactly what will be the physical, economic, geopolitical and social impacts of the phenomenon" (Chevalier 2009: 1). The following sections examine how some of these uncertainties shape petroleum controversies in Norway.

The ambiguous oil

One of my main mistakes when I began this inquiry was that I thought I was going to get a clear and coherent account of the Norwegian oil industry and the related challenges of possible future petroleum activity in the Lofoten, Vesterålen, and Senja (LoVeSe) area. Well, I did not. What I got were more like "bits and pieces" of oil bundled together in a variety of ways depending partially on the informants' roles and positions, but first and foremost on the contexts into which the oil was woven. Behind the discourse and storylines about the Norwegian "petroleum fairytale",[1] there were both contradictions and ambivalence. In a way, I was caught up in these contradictions myself as I tried to organize answers, statements, and document excerpts into neat and coherent logics before finally realizing that several parallel logics operated simultaneously in the data. A pattern finally emerged as I went through codes and coded segments several times, looking for possible links that stood out when mapping the various categories. It was not the expected pattern with clearly marked positions and perspectives but instead a pattern of multiplicity. In short, oil was many different things at the same time: Oil was ambiguous. According to Mol, multiple objects do not necessarily fall apart or imply a fragmentary character. On the contrary, they "tend to hang together somehow" (2002: 5). Drawing together the diversity of objects requires several modes of coordination (Mol 2002: 84). The focus of the analysis in this section will, therefore, be on how oil as multiple is arranged and coordinated.

I realized when going through the informants' accounts that they presented oil not only as multiple realities but also as an object that shapes and produces these various realities. According to Latour, by entering accounts, objects can be accounted for (2005: 79). Their "story" of associations reveals how their agency is being produced. This is how we can trace their connections and various links. If there are no visible effects, "they remain silent and are no longer actors: they remain, literally, unaccountable" (Latour 2005: 79). From this perspective, oil becomes a powerful organizing agent of the Norwegian economy as it participates in the government's revenues, GDP, exports, investments, taxation, welfare services, etc. What makes oil an actor or agent is that it makes a difference to a variety of institutions, practices, and polity areas of Norwegian society by changing *the state of affairs* (Latour 2005). Without an important petroleum industry, Norway would be a different society in many aspects. The following statement by a representative from an environmental organization is an example of how oil and gas are involved in both the production and the enactment of the welfare state:

Our whole social welfare system, the fact that we have the social security system we have, the good health care system, education, and such, is all thanks to our petroleum fund and that we actually found oil. We have built our wealth, our prosperity, on oil and gas resources, right, and then many people think, okay, but what is the alternative? The alternative is not going back to how it was before. Norway was actually a poor country, so what now? I believe it is difficult to envision other industries that can generate the same economic value as the oil and gas industry today. It is difficult to believe that anything can compete economically. In addition, there are many jobs within the oil and gas sector and when you do not have anything that can automatically take over, and that is real, and just say this is where I will work. That is why we say that we have to work in a parallel way. Nobody believes that you can turn off the oil and gas industry tomorrow.

As the previous example indicates, oil is also characterized by a high degree of contingency depending on the economic, political, and environmental circumstances. In the accounts of Norwegian petroleum production and possible risks provided by the informants, oil moves or transits between different logics that are sometimes linked to local circumstances and sometimes related to national or global scenarios. However, all levels can be present simultaneously in the same argument. When this is the case, one level has the ability to expose the uncertainties tied to another one. Although oil can be a certain and solid object at one level, it can become uncertain and problematic at a different one. How uncertainties link different levels, in this case, the national and the global, can be observed in the excerpt below from a senior advisor in a government agency when asked if Norway managed to stand by the political decision from the 1970s of putting a cap on petroleum production:

No, we never managed to do so. It lasted only the first 10 years or so, but at the end of the eighties, it just exploded. It went completely through the roof and there are those who think that this was … well, it was in the eighties that James Hansen from NASA came and told the American Congress about climate change. This had partially been known before, but now it became known internationally that the world had a hell of a problem, right, and then you think okay, we sit here. We have plenty of oil resources and at the same time, it seems that extracting and burning those resources will create a world which is so heated that it will become a catastrophe. What are we going to do? What if we extract all of it immediately, while it is still profitable, while there is still a market and before we reach a global agreement on a carbon tax, which means it will not be profitable anymore. That is what Norway has done. We have been on peak all this time.

The statement indicates a critical position towards what the informant believes to be Norway's position in the controversy revolving around the future of oil producers in a context of global warming and climate change: Accelerate the pace of extraction before the petroleum assets drop in value. In the early 1970s, when oil was first discovered, Norwegian authorities adopted a policy of a moderate rate of extraction to prevent the possible negative effects a substantial capital flow could have on the economy. Ryggvik and Kristoffersen (2015) highlight that this was before the climate effects of fossil fuels became scientifically acknowledged as a problem. Norway, however, while being internationally recognized as a country with high environmental standards, did not continue with its policy of moderate pace as could have been expected. "Instead, the industry and its supporters responded with a set of arguments and policies that managed to increase production at the same time as climate change mitigation became part of Norway's state agenda" (Ryggvik and Kristoffersen 2015: 250). The informant's previous statement, however, clearly shows how oil at the national level became uncertain and problematic by the international level as climate change was increasingly recognized and framed as an issue by science. Thus, before climate change was scientifically proven and, thereby, became a global political problem that had to be addressed through international climate negotiations, Norwegian authorities recognized the petroleum threat primarily as domestic. Consequently, political measures were taken so as not to heat up the economy with revenue from the newly discovered petroleum resources. Environmental risks, as we will see later in this chapter, were also national challenges that became governable through technology and new technological solutions.

The next example illustrates the political complexities that Norway faces as a petrostate with its economy and welfare system largely based on revenue from oil and gas exports. Whereas the informant emphasizes that oil is not a prerequisite or a necessary condition for having a well-functioning welfare state, his answer also sets the focus on how certainties at a national level are progressively eroding as international climate negotiations receive more attention and a climate agreement becomes more pressing:

> The welfare state developed in a period when we did not have oil revenues, when people did not believe we were going to have oil revenues. Actually, we won the lottery when we found all these oil resources and we have managed the income in a smart way through the petroleum fund. It is possible to have a welfare state without oil revenues, but if we now are going to rely on oil exploration and keep investing many of our resources, both human resources, our smartest brains, and money in the oil industry, then we need the international climate negotiations to fail. We also depend on oil prices going up as well as increased demand. But we do not want the climate negotiations to fail, and if they do not fail but succeed, then we are investing time and money in something that will not give us considerable income in the years to come. The oil industry will increasingly become an economic loss, which implies expenses for the state.

The argument reflects how oil is transformed into an object that is becoming increasingly unstable and problematic. Moreover, climate change is moving oil from being exclusively an economic asset whose biggest threat was believed to be physical scarcity towards becoming an economic and environmental threat in its own right. This could be the situation if petroleum keeps on occupying a privileged position as the engine of the Norwegian economy in a world scenario where climate negotiations have led to a common agreement. What we refer to when we talk about oil as an object is, of course, its materiality but, more importantly, it is "a crafted version of condensed presence that takes the form of a process or entity deriving from and re-enacting an ordered form of absence" (Law 2004: 162). Accordingly, crude extracted from the Norwegian continental shelf by using advanced offshore technology is different from oil translated as revenue in the national budget, and it is also different from oil as CO_2 emissions in the various carbon budgets that have been elaborated by international climate organizations. In all these versions of oil, something is included and, thereby, present, while something is also absent or relegated to the outskirts or background of the process. The connections to these absences are always a possibility and can, at any moment, be activated in discourse and policy practices. Accordingly, absence always exists *somehow* as a possibility of presence *somewhere* along the translation chain.

When going through the data, what stands out is that there are two ways of understanding the significance and the role of the petroleum industry in Norway. I have chosen to name these two accounts *safety* and *insecurity*. The reason why I landed on safety instead of security is that I believe it better conveys the meaning of Norwegian *trygghet* (noun) and *trygt* (adverb), which were terms sometimes used by informants when talking about the oil industry's role and why it is difficult to perform a *green shift* and reorient the economy to become less fossil fuel dependent. Nevertheless, the concept of *trygghet* is not necessarily unproblematic, as it does not have an exact translation and overlaps with both safety and security in English (Eriksen 2006: 11–12). "Security" could probably, therefore, have been an equally valid option.

These two versions of oil – as *safety* and as *insecurity* – with their corresponding storytelling arrange oil as two distinct objects. Although presented and recognized as different, they also appear as related, and their interconnectedness provides several converging points where safety and insecurity meet. Interestingly, what in the collective imaginary is understood as factors, which when combined form a national *safety net*, is largely what also produces oil as insecurity. Accordingly, the accounts present not only two different perspectives regarding oil production and oil dependency simply based on party affiliation or the degree of environmental commitment of the informants, but also – and this is crucial – oil is presented and evoked as multiple objects with different political, environmental, and economic "shapes". Oil is by no means passive; on the contrary, it moves around in a variety of network translations. After five decades of petroleum extraction in the North Sea, oil has become a powerful agent in Norway, which also plays an important role in organizing its own enactment. This process takes place in the assemblage of government revenues, legal framework, oil policies, petroleum taxation, investments, welfare services, and carbon taxes, etc. In other words, in the web of complex and extended relations, this is where oil becomes oil in all its multiplicity.

Safety and insecurity as modes of ordering

Following the argument from the previous section, petroleum constitutes, as we have seen, two distinct but, nevertheless, interconnected realities that are *performed* in what I will refer to as *modes of ordering* (Law 1994). Law argues that these modes are much more than simply narratives if, by narratives, we mean stories that do not work or perform something beyond themselves. The reason is that they are also, to a certain extent, "performed or embodied in a concrete, non-verbal, manner in the network of relations" (Law 1994: 20). According to Law, these modes also generate

> fairly regular patterns that may be usefully imputed for certain purposes to the recursive networks of the social. In other words, they are recurring patterns embodied within, witnessed by, generated in and reproduced as part of the ordering of human and non-human relations.
>
> (Law 1994: 83; emphasis removed)

While I address two ordering modes, *safety* and *insecurity*, that I encountered when analyzing the data, these should not be regarded as the only possible or existing ordering modes in Norway's current petroleum debate; neither should they be thought of as exhaustive nor mutually exclusive, as they have more of a hybrid character.

Consequently, modes of ordering should, therefore, be thought of as dependent variables that are both open and incomplete in the sense that they can suffer modifications depending on the intertwined contexts, as well as projected scenarios and outcomes. In other words, safety and insecurity as modes of ordering are not necessarily fixed or stable. On the contrary, I believe they are *transient*: They are temporary, they move, and they transform, but most importantly, they have a specific locality or local anchorage. If this were not the case, ordering would have been a difficult, if not an impossible, task. Law claims that ordering modes "may be *usefully* imputed to the patterns of the social for certain purposes" (1994: 84). Consequently, different modes of ordering perform and arrange different things or different realities. Even if they could turn out to have a more general applicability, they are always located somewhere within a specific performance or enactment. The temporal and spatial dimensions of ordering modes are highly relevant if we take into consideration what I have already said regarding how safety at one level can transform into insecurity at another depending on a wide array of circumstances. The analysis that follows is, therefore, specifically related to how safety and insecurity work as ordering modes within the Norwegian petroleum debate, informed, among other things, by low oil prices and a depressed oil market on the one hand, and uncertainties and expectations regarding the outcome of the Climate Summit in Paris in December 2015 on the other.

The reason why some factors are involved in ordering safety, whereas others are related to the ordering of insecurity, has to do with the interplay between global, national, and local levels. These levels should not be taken as points of departure or as given, fixed conditions that offer some sort of causal explanation (as when an event is explained by a surrounding context) but as effects of networks of the social. The global is located at many different sites at the same time: This is what really makes a phenomenon global (Latour 2005). Through extended networks, localities can connect and thereby "nationalize" or "globalize" themselves. Size and scale are, therefore, effects of patterned performance that have stabilized in one way or another. According to Callon and Latour, size is specifically a matter of "blackboxing"[2], as some actors manage to grow by the solidity or durability of their materials and by how these are hierarchically arranged and locked into black boxes. Accordingly, the analysis of micro- and macro-actors is not simply a matter of size (Callon and Latour 1981: 284–285). We need to examine carefully "the processes by which an actor creates lasting asymmetries" (Callon and Latour 1981: 286). Consequently, the petroleum industry has managed to become a macro-agent of the Norwegian economy by enrolling an ever-increasing number of entities from a variety of fields (investments, materials, labor, research, technology, etc.), often at the expense of other sectors and industries, which are unable to compete in the same conditions.

Safety tells and produces storylines about what makes petroleum *safe* and, more specifically, it tells how the Norwegian petroleum industry provides the citizens with safety and work as a shield against the unpleasant state of uncertainty and risk experienced in other parts of the world. The discovery of oil in the North Sea constitutes a *turning point* in the country's economic history. In 1969, Phillips found oil on the Norwegian shelf, and a few months later, in 1970, it was confirmed that the Ekofisk field contained huge amounts of petroleum resources (Ryggvik 2015: 8). By establishing oil as the turning point, this mode of ordering explains how the discovery of oil changed the country's direction: There is a "before and after" oil. Accordingly, the mode of *safety* arranges Norway as a safe place to live largely thanks to the oil wealth and smart people, the *pioneers*, who developed the oil industry in such a way that the oil wealth came to benefit the majority of the population. It thereby highlights that this is not always the case in other oil-rich nations, where oil is more uncertain and politically problematic, as it has often only enriched financial and political elites or foreign oil companies and not the local population. Regarding this situation a politician from the Labour Party commented:

> In comparison with perhaps other oil nations … Norwegians, generally speaking, have been able to take part in the oil wealth. There has been a politics of distribution, which has made it possible for the whole population to participate in the "petroleum fairytale", if we can call it that. It was a poor population and the social democratic politics has made it possible for people to be a part of it through the welfare state.

The dualism at work here is the *inclusive oil*, the one that democratically reaches the whole population, versus the *exclusive oil* that only benefits a few privileged sectors of society. The former tends to have high social and environmental standards as well as safety measures, while the latter is frequently associated with resource depletion, environmental degradation, and a lack of safety measures for the oil workers. Consequently, this mode of ordering performs the oil industry as a set of relations that positively provides the population with welfare and a considerable degree of economic stability. However, it has not been easy or straightforward by any means. On the contrary, it has been a long learning process with high stakes. Therefore, a lot of telling is about risk, political decision-making, firm planning, vision, and precaution. A lot of precaution is actually scattered around in this mode of ordering, at least when the beginning of the oil adventure is presented. An illustrative example is white paper 25 (1973–1974), which states:

> The petroleum discoveries in the North Sea will make us a wealthier nation. The government believes that the new opportunities should primarily be used to develop *a qualitatively better society*. It is important to avoid outcomes that are reduced to rapid and uncontrollable growth in the employment of material resources while society remains without substantial transformation.[3]
>
> (Finansdepartementet 1974: 6)

The white paper's following paragraph emphasizes the importance of the new economic opportunities that the oil industry has the capacity to create. The idea is to generate equal living standards and prevent social inequalities. There is also a focus on the importance of developing a production system that is environmentally friendly in its resource use. Another central point is to develop the welfare state further and the creation of employment, especially for those groups that have difficulty entering the labor market (Finansdepartementet 1974). What is being produced is *welfare and safety for the collective as a whole*, not sudden wealth as the result of the ambition of a privileged group or an exclusive sector of society. Consequently, the task is to develop a strong industry, which means designing political strategies to manage the petroleum resources properly and make them last into the future.

Moreover, safety is a mode of ordering that informs us about an industry that was born and shaped by legal and political decision-making. The result was frameworks that oriented the incipient new oil industry in a specific direction. The frameworks shaped and produced what would become not only Norway's biggest industrial sector but also a significant service and supply industry. In other words, the fact that the oil industry became an economic and industrial success was not just a coincidence nor a stroke of luck. It was the result of carefully designed political and legal mechanisms that put policies, knowledge, and technology to work. The principles that guided the process can be observed in various documents that were produced during the early years of the petroleum industry. Accordingly, they constitute important instruments and tools used to create a solid foundation to secure national control over the petroleum industry. One example is the aforementioned white paper 25 from 1974, but other principles contained in other texts are also highly relevant. This is particularly the case in what came to be known as the *ten oil commandments* produced by the Norwegian parliament's committee on industry in 1971:

1. National supervision and control must be ensured for all operations on the Norwegian Continental Shelf (NCS).
2. Petroleum discoveries must be exploited in a way which makes Norway as independent as possible of others for its supplies of crude oil.
3. New industry will be developed on the basis of petroleum.
4. The development of an oil industry must take necessary account of existing industrial activities and the protection of nature and the environment.

5. Flaring of exploitable gas on the NCS must not be accepted except during brief periods of testing.
6. Petroleum from the NCS must as a general rule be landed in Norway, except in those cases where socio-political considerations dictate a different solution.
7. The state must become involved at all appropriate levels and contribute to a coordination of Norwegian interests in Norway's petroleum industry as well as the creation of an integrated oil community which sets its sights both nationally and internationally.
8. A state oil company will be established which can look after the government's commercial interests and pursue appropriate collaboration with domestic and foreign oil interests.
9. A pattern of activities must be selected north of the 62nd parallel, which reflects the special socio-political conditions prevailing in that part of the country.
10. Large Norwegian petroleum discoveries could present new tasks for Norway's foreign policy. (Norwegian Petroleum Directorate 2010: 1-4)

While the majority of these principles have governed and oriented the petroleum sector for decades, others suffered modifications as political and economic realities and opportunities changed. Nevertheless, they bear an important testimony of the political processes and institutional practices that made the new industry succeed and, therefore, constitute an essential part of what we could call the *origin story*. As safety oil is not only entangled in politics, technological innovation, economy, and welfare, it is also performed as something with a history. It has managed to become an iconic industry that defines the country just like a tradition: Norway is an oil nation and Norway knows what there is to know about the oil industry. In this context, names and symbols constitute important elements of a country's cultural heritage and are, therefore, cleverly used by the Ministry of Petroleum and Energy. Regarding its naming practice, the Norwegian government states:

> The names of many fields in Norway are taken from Norse mythology, with strong roots and steeped in national tradition. This is a tradition that should be continued. However, the strongest names from Norse mythology are already in use, which means that we should also consider new types of names. The names given to larger fields in new areas should reflect the industry's importance, both for specific regions and for the nation as a whole. The Ministry therefore plans to make adjustments in the naming of petroleum deposits, to ensure that they fit into a national context and history.
>
> (Norwegian Ministry of Petroleum and Energy 2011: 7)

Consequently, safety as an ordering mode reifies the petroleum industry as a known, conquered, and, by now, "traditional" object. In other words, oil has become safe by expert knowledge and technological innovation. The mode of safety coordinates the Norwegian petroleum industry's past and present by tying together and highlighting specific events and bringing them to the forefront while others become marginalized or silent. It explains how things began and why they ended up being the way they are today. In short, it explains how Norway went from being an inexperienced actor in the production of hydrocarbons to becoming a "hydrocarbon superpower".[4]

As stated by the Norwegian Petroleum Directorate, when referring to the oil commandments: "These principles have subsequently been significant for the direction and shape of the Norwegian petroleum policy" (2010: 1). They are part of the petroleum sector's origin story, which performs the early years as a learning process where Norway started from scratch. Reasonable politicians represented by a handful of visionaries, the pioneers who entered the unknown territory of petroleum extraction, understood that in order to master the oil industry and create an industry for the future, they had to learn how "to play the game", which implied learning from those with knowledge: The international oil companies. The important thing was to gain the knowledge and technological competence needed to master the complexities of the industry and secure the future national management of the petroleum sector. Law maintains that stories of vision can also include the process of apprenticeship during which a series of *rites of passage* play a decisive role (1994: 80). This is very much the case with safety as a mode of ordering: The incipient industry had to learn step by step before reaching the stage of a mature and autonomous petroleum sector. Accordingly, the new industry had to learn how to solve technological challenges in order to "secure positions at each stage of the oil process, from upstream exploration and production to refining, the chemical industry, and the sale of oil products" (Ryggvik 2015: 10). However, Norway managed to assimilate all sides of the industry through hands-on learning, and the country never lost national control over the management of the petroleum resources. This had a lot to do with the creation of Statoil (now Equinor), the state-owned oil company. Although an extremely important actor that has contributed enormously in making Norway one of the wealthiest countries in the world, the company is far from being the real hero in this ordering mode. Similarly, politicians, engineers, and economists were important in the early years of the petroleum sector, and they still are due to their foresight, planning, and innovative capacity. Nevertheless, while being important agents in the storytelling about how oil became a national safety net, they participated in the creation of even bigger agents, the real "heroes" of this mode of ordering: The Petroleum Fund and the offshore-related service and supply industry. In other words, they are an economic and technological hero. These are the real subjects in this ordering mode as they provide safety at present and are expected to provide safety also for future generations. The Petroleum Fund, which currently constitutes one of the biggest investment funds of its kind, reached a total value of 10,000 billion Norwegian crowns in 2019.[5] Conversely, the innovative technological community that rose to the challenges of the harsh conditions of the North Sea and produced cutting-edge technology for deep waters is a powerful actor-network that is expected to adapt and, thereby, respond to the new technological challenges of the green shift by using its experience and technological skills.

Insecurity is a quite different mode of ordering. The term, however, could make us believe that this ordering mode is somehow the opposite of *safety*, but this conclusion would be misleading. A more appropriate way to see it is that *safety* as a mode of ordering is involved in how petroleum issues are framed and defined by *insecurity*. Specifically, it tells how some of *safety's* agents have become increasingly insecure. In other words, there are materials, markets, actors, and relations that have been solid and stable in the past and that are currently "on the move". Things are not as they used to be in the petroleum sector as a member of the People's Action for an Oil-Free Lofoten, Vesterålen, and Senja emphasized in an interview:

> We see that these studies, oil-engineer studies and things like that, have considerably fewer applicants than before. I think that older people still experience oil as being something new, something that is going to give us many different things in the future, but young people realize that if they are studying to become oil engineers, there is no guarantee that they will get a job now while they are young, or have a job in 20 or 30 years. So, it is clear that young people see the future in a different light and they are probably more aware of the environmental part because they have grown up with these things.

As we can see from these reflections, the oil industry is no longer considered a sector that can guarantee future job opportunities, which makes young people seek other career options, something that numbers from the University of Stavanger clearly indicate. There was a 50% to 70% decline in applications to oil-related engineering studies from 2015 to 2016. Other Norwegian universities that offer petroleum studies have also experienced the same trend (Norli 2016).

Besides the low oil prices in recent years, climate change is also playing an important role in transforming oil into a less solid and coherent object. In other words, the climate issue is arranging the national petroleum industry as a series of political and economic uncertainties, which are producing ambivalent decision-making processes. Commitments to reduce greenhouse gas (GHG) emissions and eventually to become a low-emissions society in 2050 are part of the government's plan to carry out the green shift at the national level. What this green shift implies seems unclear in many aspects. While the Ministry of Climate and Environment (2014) frames the transition as located primarily in transportation, housing, agriculture, landuse, industry, and business, there is silence when it comes to how this transition towards a low-carbon society will play out in the oil and gas industry. Although the commercialization of hydrocarbons works as the engine of the economy, the petroleum and gas industry is strikingly absent as a "site" where the green shift will be performed. The absence and exclusion of oil make it highly uncertain as its role in the transition is being enacted as a matter of a completely different rationale that belongs somewhere else; the question is where? The Ministry of Climate and Environment states that the goals for the green shift, which are based on the government platform, are, among others, to achieve "a changed economic dynamic or growth potential, a 'new economy' that is itself arranged on solutions that produce low greenhouse gas emissions (New Climate Economy), and reduces the need for subsequent remedial environmental measures" (Ministry of Climate and Environment 2014). Although the purpose is to change the dynamic of the present economic model, oil as the leading sector in value creation is not mentioned, nor does the text specify what this "new economy" actually is. This mode of ordering exhibits what Law (1994) refers to as a pattern of *deletion*. Consequently, *insecurity* tells about how oil is not present in the planned green shift but relegated somewhere else. It is consistently being secluded and made invisible as it travels between conflicting logics. However, on other occasions, petroleum is heavily present, but outside the storylines of the green shift: In January 2015, the Ministry of Petroleum and Energy launched the 23rd licensing round. Fifty-four of the announced blocks were in the Barents Sea. According to the exploration director in the Petroleum Directorate, the licensing of new blocks located in unexplored areas could on the one hand "open up a new petroleum province in the Barents Sea", and on the other, it "helps keep the activity level high" (Norwegian Petroleum Directorate 2015). Hence, this ordering mode works to address the breach between the official discourse and statements about a transition towards a green, low-carbon economy and the dynamics of the petroleum sector working to perpetuate itself into the future. In recent years an unstable market has made oil a less secure object. In addition, climate change and pressure from some political parties and civil society groups on the government to initiate a transition and phase out fossil fuels are also contributing to this instability. However, these factors are only a part of the story. This mode of ordering tries to display something else: *Insecurity* inhabits *safety* as a future possibility. As a latent possibility, it also erases the clear-cut boundaries of oil as a conquered and mastered object. As previously mentioned, objects are condensed versions of something: Certain materials, relations, inscriptions, and actors are included and thereby present while others by exclusion have become marginalized into a state of absence. By coordinating some of the absences of *safety*, *insecurity* is brought to life.

Whereas low oil prices and climate change problematize oil as the backbone of the Norwegian economy, it is principally *the safety and stability of oil* that transforms it into *insecurity* in this mode. The very same process that produces safety also generates insecurity because it slows down the transition towards a carbon-constrained economy and the green shift. In other words, the oil wealth has been and still is the solution to many problems. It is therefore challenging to reorient the economy. This is often referred to as path dependence in economic and institutional theory. A resource, technology, institutional, or economic model once established is not easy to change or modify, as they tend to stabilize. On some occasions, small events can determine a solution that, once established, tends to lead down a specific path (North 1990: 94). Later on, it is economically or politically too costly to change direction or select a different alternative. Similarly, to perform the green shift and transit towards a less oil-dependent economy is difficult, first and foremost, because it has been a success story. The oil wealth has given the population welfare services and economic growth. In short, it has provided safety, and this is what makes a green transformation difficult. An informant explained the challenge as follows:

> There has been an incredible income, but there is also another popular term, the green shift, that many people talk about, but not everybody wants to put so much into. I think it is also a bit scary. There is no doubt that there have been some crazy revenues from oil and we have the Petroleum Fund, right. We live so safe in Norway that there is almost nothing that can hurt us, but at the same time, we know that we will run empty. We know that we cannot keep on extracting and that we have to do something else. It does not need to be scary because we have a lot of resources.

How extremely difficult it is to phase out an industry that has been so profitable and done so much for the Norwegian economy and labor market is evident also in the following statement that emphasizes the ambiguity of oil:

> There is a reason why we do not manage to produce the green shift, and it has probably been easier to work on elsewhere because we have been lucky for having oil and we have been unlucky for having oil, because it means that there has not been any pressure to change things.

These two modes of ordering, safety and insecurity, collide since they arrange oil as two different objects by organizing present circumstances differently. My argument is that this situation takes place in the intersection between *oil as safety* and *oil as insecurity*. This is where the oil economy is enacted as uncertain; nevertheless, the two ordering modes are closely interconnected. Insecurity as a mode of ordering organizes oil by "weaving" it into international and global challenges, primarily the oil crisis and climate change. In this way, oil increasingly becomes a matter of concern. However, the contexts (the international oil market and climate change) are not something exterior that surrounds the issues at hand "but something which is integral to the very action" (Asdal 2012: 388) and, thereby, makes ordering possible. This is how oil takes on different economic, political, and environmental "shapes" and becomes multiple.

Policies + knowledge + technology = mastering the oil industry

Safety as a mode of ordering is largely concerned with how Norway came to master the oil industry. When I looked at the accounts about what informants believed to be the outcome of more than 50 years of petroleum activity, there were certain elements, entities, and actors that repeatedly seemed to be performing the *state of mastering*. In other words, they were working to give the oil industry this specific form. Consequently, I realized that I needed to take a closer look at *what* and *who* cooperated in order to produce oil as a national safety net, despite its contingent character. This section is, therefore, primarily dedicated to the analysis of how different associations and network relations synergistically came to produce a strong and competitive petroleum sector. In order to trace the different elements at work, I examine what the accounts highlight as important achievements since Norway first started its petroleum production in the early 1970s. The following statements are representative segments of these accounts:

> The way we built up and organized the petroleum industry in Norway by establishing a state-owned oil company was really smart. By doing so, we secured a huge share of the revenues for the treasury and we also established the Petroleum Fund, which makes it possible to manage the money in a more long-term perspective. Additionally, we have acquired a lot of competence that we can use in other industries such as offshore, wind power, etc. We can learn a lot from the way we have organized the oil industry.

This segment emphasizes the importance of policies that led to the creation of Statoil, which became an important actor in the governance system of the petroleum industry in Norway. The state was also able to secure a considerable share of the revenues by securing national control over the oil resources. Additionally, the establishment of the Petroleum Fund not only links Norway's contemporary inhabitants to the welfare state, but future generations are also expected to benefit from the oil wealth. Along the same line, the next statement addresses the importance of laying the foundations for a national industry instead of applying "weak" oil policies that primarily benefit well-established international oil companies and not the country that owns the resources. As stated by a representative of the County Governor of Nordland (in northern Norway), this has, sadly, been the case in many incipient oil-producing nations worldwide:

> In many countries where you have oil extraction, the majority of the revenues have gone to those who produce the oil. Oil may not have gone into a major social context. When Norway opened up to petroleum production, a proper national control was secured so that Norway as a nation would get a substantial share of the oil revenues. This was done early on and if this had not been the case, there was the risk of Norway becoming, politically, an oil province for Esso, BP, Shell and these [companies].

While the two previous statements refer to Statoil, the Petroleum Fund, and national control over petroleum resources as part of mastering the oil industry, the next statement touches on another important political tool: The concession regime for the allocation of oil blocks. It specifically emphasizes the importance of shared rights between the state-owned company and international oil companies when production licenses were handed out. In the early years, the licensing rounds were key opportunities for Statoil to learn "the trade". Consequently, to gain knowledge of how to operate the oilfields, shared ownership was an important instrument for competence building and the development of technological skills.

Regarding these matters a representative of the County Governor of Nordland commented:

> Norway made demands concerning those who obtained oil concessions: The concessions should be [developed] together with what was originally called Norol and that later became Statoil, in order to guarantee that Norway had a considerable part of the ownership and that Norwegian oil companies had the opportunity to build their own competence and eventually manage the industry in the future.

Other statements bring knowledge to the forefront as one of the most important achievements after five decades of petroleum production. In the accounts of how oil is becoming less stable, competence-development and knowledge as forms of capital that can be invested in other industries or areas of society are frequently emphasized. Another significant aspect is how oil revenues have directly financed education and research as highlighted by an advisor at NHO (The Confederation of Norwegian Enterprise):

> If we ignore the fact that we have physically "eaten" many of the results, we have evidently reached a high level of knowledge, especially knowledge about offshore operations, which we can benefit from within ship designing and security at sea, within offshore wind farms, well within many fields. We have a big export-oriented oil service industry and, of course, it has complicated days right now, but that has also happened before. The oil revenue, both directly and indirectly, is the major contributor to the production of knowledge at colleges, universities, and within research. These are partly financed through the national budget and partly through direct [financing].

As we can observe from the previous excerpts, there is clearly interrelatedness and interaction between policies, knowledge, and technological innovation in the enactment of the *state of mastering*. These fields seem to operate and work together in conquering and stabilizing oil. We need to find places where this interaction can be observed to understand how these fields produce oil as a specific kind of object. Accordingly, important construction sites (Latour 2005) are policy documents that gave the current oil industry its shape and orientation. These kinds of texts are important since they mobilize a series of principles, resources, and actors in order to distribute roles and responsibilities. The aforementioned white paper 25 (1973–1974), therefore, constitutes a relevant location where the "birth" and enactment of the Norwegian oil industry can be closely examined. As Asdal argues, documents are not mere descriptions of an external reality; they shape, transform, and modify that reality (2015: 74). Documents are important, as they give us access to how institutional practices and objects are performed. Policy and planning documents have a particular potential for agency in this context, as they formulate objectives and desired outcomes (Kaarhus 1999: 34). In other words, materiality is "made available to us by way of documents – that is, in a material-semiotic version" (Asdal 2015: 75). Consequently, documents should not be regarded as something separate or autonomous from material reality since they constitute part of that very same reality. As Law has pointed out, "words and worlds go together" (2004: 33). This means that, in relation to oil, documents such as white paper 25 actively work to produce oil as a durable and stable object by defining its relations, fields of action, roles, restrictions, and possibilities.

When we examine this particular white paper, it becomes clear that mastering the complexities of oil has a lot to do with carefully planning the new industry's present and future. However, the document also enacts oil as moving or traveling between certainty and uncertainty, between possibilities and risks. How to handle some of oil's contradictions is, therefore, an important objective present in the text. The title, *Petroleum Activity and its Position in the Norwegian Society*,[6] is literally opening up a space and locating the industry by defining its various fields of action. While issued by the Ministry of Finance and written as a series of economic recommendations, calculations, and possible trajectories, it is not exclusively an economic document. At first sight, the text seems to perform petroleum primarily as an economic issue where both possibilities and problems related to the industry are analyzed and defined. However, while being an economic issue, petroleum also entails important and decisive societal aspects that the text introduces as part of a democratic practice and discourse:

> Through this whitepaper, the government wants to lay the foundations for a broad debate on oil policies among the Norwegian people at large. Great importance is therefore given to clearly convey the tasks and problems we face and the possibilities we have.
>
> (Finansdepartementet 1974: 5)

By inviting the Norwegian people to an inclusive national debate on the place and future of the petroleum industry, the document frames oil as a national object with ramifications deep into various fields of society. Additionally, the text's introduction contributes to the enactment of oil as a contingent reality by explaining that things are by no means settled or stable, as new information could, in principle, change the present character and situation of the industry: "New information surfaces continually, which could make it necessary to review the issues at hand. Evidently we may have to perform a new broad assessment of the whole petroleum industry" (Finansdepartementet 1974: 5).

The document performs the uncertain and contingent character of oil in various ways, not only by constantly referring to the new economic and social possibilities oil is expected to create but also by emphasizing the problems Norway could face due to the sudden oil wealth. However, other sections of the document work to counteract these uncertainties by contributing to the solidity and durability of the petroleum industry. In other words, a lot of *stabilizing work* is performed in order to add to oil's certainty. By a series of calculations, oil is translated into reserves, exploration trajectories, and extraction rates. The text establishes that existing petroleum discoveries would secure production through the 1990s. It calculates how much oil the proven reserves contain and how much oil will make up the exportable surplus in the 1980s. It tries to establish when production will most certainly decline in already operating blocks (when they will peak) and when national production is due to fall if new oilfields are not discovered. Alternatively, if new reserves are added to the current ones by keeping up exploration as it is at present, there is the potential to double the country's petroleum production.

These are just a few examples of how uncertainties are approached by numeric translations. In similar ways, the document contains calculations and trajectories of public expenditure, investments, domestic consumption, and transformations of the labor market, etc. All these calculations and economic projections add to oil's stability. By translating oil into numbers, it becomes tangible, less abstract, and increasingly governable. Calculation systems and economic projections are, therefore, important instruments to enact oil, despite all its uncertainties, as a planned and governable object: Oil is made accountable.

Moreover, not only is oil being translated into oil revenues in national and long-term budgets. Oil itself is performed as a possible mediator that inherently has the capacity to transform Norwegian society. In other words, oil is not described simply as an intermediary that will take the country from A to B, which would be reducing its performance to the act of *transportation* (Latour 2005). Hence, Norway will not only go from being a country with high living standards to having even better living standards. On the contrary, if managed properly, the oil revenue will produce not only a wealthier nation but also a "qualitatively better society" (Finansdepartementet 1974: 6). Consequently, oil is enacted as a mediator with the capacity to translate and, thereby, qualitatively transform Norway as a nation.

The document can be read as a series of guidelines on how to perform this transformation. It is a sort of handbook, in other words, on how to use oil to carry out an improvement of current societal conditions. The text can also be read as a constitutional document. Similar to a political constitution that establishes the foundational principles on which a nation rests, white paper 25 displays the principles, organizational structure, and the distribution of competences of the oil nation. The document provides an opportunity, therefore, to study a petrostate in the making by looking into the principles that will govern the new petroleum industry. As we know, no constitution comes without a dream. The dream is actually twofold in white paper 25: The first part is a qualitatively better society that can be achieved by carefully managing the resources: "The guidelines, which have been outlined for the petroleum industry and the use of revenues, should be part of a planned transformation of the Norwegian society" (Finansdepartementet 1974: 6).

The second part is to create a leading Norwegian petroleum industry based on the participation of local actors:

> If we are to achieve a broader Norwegian petroleum involvement and keep up with, and maybe lead, the development in certain fields in the years to come, an extensive research and development program is required. However, further research and assessment activities should involve the

development of society at large and not only be organized specifically around the petroleum activity.

(Finansdepartementet 1974: 10)

The activity in the North Sea represents the first large scale petroleum exploration and extraction in harsh weather conditions on depths of more than 70 meters. It is decisive for the Norwegian industry to participate in the development of new technology for exploration and extraction in these waters. Active Norwegian involvement in this phase will secure participation also in the future.

(Finansdepartementet 1974: 18–19)

As these excerpts indicate, a strong national involvement is framed as a matter of knowledge, technological development, and know-how. Norway must develop research programs that are directly related not only to the petroleum sector but also to social development, in general, to gain competitiveness in this new industry. However, research on "society at large" is apparently understood as part of a social planning process required to comprehend and control transformations in work, industry, and settlement due to oil (Finansdepartementet 1974: 10). The document also emphasizes the importance of both innovative thinking in several fields and the capacity for using existing knowledge. By addressing future growth in the petroleum industry as a matter of knowledge and research development, oil is turned into an epistemic object. As previously mentioned, oil is accompanied by several uncertainties, and one way to stabilize the resource is through economic calculations. Another way to make oil solid is through research and scientific knowledge. By transforming oil through modifying work (Asdal 2015), the document is adding to its complexity, as oil is not solely an exportable energy commodity but is also a particular kind of knowledge object. Knorr Cetina (2001: 181) defines these objects "in terms of a lack in completeness of being that takes away much of the wholeness, solidity, and the thing-like character they have in our everyday conception". As already mentioned, the text performs oil as an object, which is both fuzzy and unfinished as new knowledge can make it increasingly uncertain and bring about the need to reevaluate the whole petroleum industry, or on the contrary, add to its certainty by layering new knowledge. In other words, objects of knowledge are objects that have not been "blackboxed" but remain open in all their incompleteness. According to Knorr Cetina (2001: 181), as they "are always in the process of being materially defined, they continually acquire new properties and change the ones they have".

Oil is a highly uncertain object with several economic, environmental, and social challenges, so *new knowledge* becomes a significant political tool that performs what I have previously referred to as stabilizing work. In the white paper, the petroleum industry acquires a privileged position as to what should become prioritized research fields. Consequently, an important part of producing a petrostate is making space for it within the research community and establishing research programs. That specifically means, in this case, that oil acquires a dominant position as a knowledge object at the expense of other uncertain objects:

> Research will be required in many fields. It goes without saying that with the resources a small country as Norway has at its disposal, especially regarding professionals, there will be a need to select some tasks and let go of others. To engage in new fields will often be at the expense of other activities that were initiated earlier in other areas. It is clear that certain research and assessment tasks should be highly prioritized.
>
> (Finansdepartementet 1974: 10)

The research fields that the document lists as priorities are all related to petroleum production in one way or another. They range from knowledge about the ocean and the continental shelf to technology for exploration and extraction and from foundations for social planning related to work and industry to petroleum as part of global resource perspectives (Finansdepartementet 1974: 10). Asymmetries are created regarding other research areas and existing industries by making petroleum and related industries dominant research fields. Consequently, the petroleum industry has been able to attract resources at the expense of other sectors from the start, enabling it to grow in size and power. Many of these asymmetries have stabilized since the early years of the oil industry, clearly exposing the entanglement of politics and scientific research. The present white paper is, therefore, an important example of how hybrids work and participate in the enactment of realities.

The document focuses on a variety of issues that could become future problems due to oil's economic, environmental, and social uncertainties. Interestingly, a particular policy instrument with multitasking purposes repeatedly shows up in the text: The moderate pace of extraction. The government feared the negative impacts of considerable oil revenue suddenly injected into the economy and, therefore, recommended a controlled rate of extraction. This is also inscribed as an appropriate way to prevent major environmental problems, as a moderate extraction rate makes things more controllable. Nevertheless, the white paper states that specific restrictions and constraints on petroleum production can become necessary in highly sensitive areas (Finansdepartementet 1974: 16). Another argument tied to the government's recommendation of applying a moderate pace is that since oil is non-renewable, it is important to extend the resource and, thereby, establish a long-term perspective for the industry. White paper 25 enacts oil as an object that is opposed to and, thereby, in competition with other resources and industries. The dualism at work is non-renewable versus renewable, or more specifically, oil versus fish:

> Fish constitutes an important food resource, which is renewed from one year to another. Petroleum is a resource that will run out. As far as possible, we have to make sure that its exploitation does not cause damage to life in the

sea. The extraction should not cause considerable obstacles or limitations for the fisheries.

(Finansdepartementet 1974: 15)

The white paper clearly envisions the possibility of a territorial conflict between the petroleum industry and the traditional fisheries, which are framed as renewable resources as opposed to non-renewable resources. This is still one of the key issues today in the Norwegian petroleum debate, particularly in the case of the LoVeSe area.

Although many additional things could have been said about the multiple roles oil is given in white paper 25, I have chosen to focus on how oil is enacted as an incomplete and unstable object that is being mastered through different mechanisms, such as political decision-making, economic translations, and new research areas. The incipient petroleum industry is performed as a series of opportunities, primarily as a vehicle to extend and further develop the welfare state but also as an instrument to create a new national industry based on technological innovation, competence building, and scientific knowledge. Conversely, oil is far from being enacted as a solid or stable object, as it inherently has the ability to generate social and economic problems, environmental risk, and weaken other existing sectors as it increasingly attracts material and human resources. Accordingly, the document enacts oil as an epistemic object that lacks "completeness of being" (Knorr Cetina 2001: 182), as it is never complete in itself but unfolds and acquires different shapes, depending on a variety of circumstances, even "on how its future develops" (Knorr Cetina 2001: 181 with reference to Rheinberger 1992). Hence, white paper 25 unpacks oil along a bifurcating route where oil moves between benefits and costs. Petroleum in itself is, therefore, neither completely certain nor uncertain, but it acquires its characteristics as it enters into associations with other entities in various fields (locally, nationally, and internationally).

Conflicting logics and the Norwegian argument

The IEA declared in its *World Energy Outlook* (*WEO*) (2012) that two-thirds of total, proven fossil-fuel reserves, estimated to amount to 2860 gigatons of carbon dioxide ($GtCO_2$) of emissions, had to remain in the ground if the world was going to achieve the 2°C goal (IEA 2012: 259). The statement received considerable attention: Not only did the report state that extraction of fossil fuels had to slow down drastically prior to 2050 but, interestingly, the alert came from a non-environmental organization.[7] However, the IEA is not the only organization that has presented this type of warning. The Carbon Tracker Initiative, a London-based think tank, has similarly elaborated several carbon budgets that clearly illustrate the urgency of implementing policies that can reverse the current trajectory of increasing CO_2 emissions. In its first report on *unburnable carbon* (2011) that covered the period 2000–2050, the total budget was calculated to be 886 $GtCO_2$, of which 282 $GtCO_2$ had already been released from the burning of fossil fuels and an additional 39 GtCO$_2$ due to land-use change (Carbon Tracker 2011: 6). This left the world with only 565 GtCO$_2$ for the next 40 years, while the global proven reserves of fossil fuels, if consumed, had the potential to release almost five times more CO$_2$ into the atmosphere (Carbon Tracker 2011: 6).

The carbon budget is considerably higher, however, in an updated analysis from 2013 developed in collaboration with the Grantham Research Institute on Climate and the Environment (London School of Economics). For an 80% probability of not exceeding a 2°C average rise in global temperature, the budget is estimated to be 900 GtCO$_2$, and for a 50% probability, 1075 GtCO$_2$ (Carbon Tracker and Grantham Research Institute 2013: 4). This budget has an important difference, however, compared to the previous one: It assumes significant reductions of other non-CO$_2$ GHG emissions towards 2050, such as methane. This means that additional reductions in other sectors such as agriculture and waste are accounted for in the extended budgets (Carbon Tracker and Grantham Research Institute 2013: 4). Nevertheless, after 2050, the quantities of burnable carbon contract significantly, leaving the world with a drastically reduced budget. Whereas carbon capture and storage (CCS) technologies could, in principle, expand the carbon budget, their implementation so far is very limited. Moreover, if we take the IEA's 2012 *WEO* report that estimates the total reserves in terms of emissions to be 2860 GtCO$_2$, not extracting two-thirds of the reserves will give us a carbon budget of around 900 GtCO$_2$ (IEA 2012: 241). This is the same as the one provided by Carbon Tracker, with the difference that in the Carbon Tracker and Grantham Research Institute 2013 report, this budget only gives an 80% probability of achieving the United Nations Framework Convention on Climate Change (UNFCCC) 2°C objective. Hence, the differences that can be observed between the various carbon budgets have to do with how the total volumes of fossil fuels are calculated. In other words, fossil-fuel reserves are not stable or fixed dimensions but depend on a wide array of circumstances that influence the estimates being used. According to the IPCC, "changing economic conditions, technological progress, and environmental policies may expand or contract the economically recoverable quantities altering the balance between future reserves and resources" (IPCC 2014a: 525). Consequently, not all oil and gas reservoirs are recoverable at present under current economic and technological conditions. However, as circumstances change, these resources have the potential to move into the category of reserves.

Furthermore, IPCC's *Fifth Assessment Report* (AR5) from 2014 declares that "emissions of CO$_2$ from fossil-fuel combustion and industrial processes contributed about 78% of the total GHG emissions increase from 1970 to 2010, with a similar percentage contribution for the increase during the period 2000–2010" (IPCC 2014b: 5). With different scenarios outlined as mitigation pathways, the document presents various degrees of probability of reaching the 2°C goal, depending on a series of variables. In a "business as usual" scenario, with no further mitigation efforts to reduce GHG emissions than those already implemented today, emissions will most certainly continue to rise, leading to an increase in average temperature ranging from 3.7°C to 4.8°C above pre-industrial levels (IPCC 2014c:

8). On the contrary, with a threshold of 450 parts per million (ppm) CO_2-eq by 2100, global warming is likely to remain below 2°C. However, this requires a drastic reduction in GHG emissions (40% to 70%) prior to 2050 (IPCC 2014c: 10). Accordingly, the Paris Agreement's expressed intention to pursue a halt to the temperature increase at around 1.5°C, evidently requires a more restrictive carbon budget than those outlined for a rise in temperature around 2°C.

Although the inclusion of different temperature thresholds, baselines, timeframes, and types of emissions (GHG emissions in general, CO_2 emissions only from burning fossil fuels, or CO_2 from fossil fuels and other sources) can make a direct comparison between the reports difficult, they all translate the relationship between human-induced GHG emissions and the rise in global average temperature. Fossil-fuel combustion plays a significant role in this relationship. When we break down the global energy mix (total primary energy supply), oil is the most important primary energy source at 31%, followed by coal at 29%, gas at 21%, and other sources at 19% (hydro, nuclear, solar, geothermal, wind, biofuels, etc.). The order changes, however, if we look at CO_2 emissions, as coal is responsible for 46% of the emissions, oil 34%, gas 19%, and other 1% (IEA 2015a: 6). These differences have to do with the carbon intensity of the unit of energy produced: With an average of 92.0 grams of CO_2 emitted per megajoule of energy produced (gCO_2/MJ), coal has a higher environmental impact than oil with 76.3 gCO_2/MJ and gas with 52.4 gCO_2/MJ. The level of CO_2 is assumed to be higher, however, in the case of unconventionals (IPCC 2007: 264). Despite differences in levels of CO_2 emissions, all fossil fuels release carbon dioxide when combusted and are, therefore, an essential part of the climate issue. Consequently, the majority of the currently known reserves are simply not extractable if the global increase in temperature is going to stay below 2°C. In other words, they constitute *unburnable carbon*. If this means that we need to keep two-thirds or four-fifths of proven fossil-fuel reserves in the ground is actually a secondary matter, the important thing is that all these expert reports point in the same direction: Our energy system is no longer sustainable and requires a set of major and complex transformations before a low-carbon energy system is in place. Another important factor that these documents have in common is the emphasis on a relatively short timeframe to perform the political, economic, technological, and social transformations required to prevent a climate catastrophe.

Against this background, it becomes clear that while the current climate regime frames climate change primarily as a "carbon emissions issue", the extraction of fossil fuels is where this issue originates and actually starts its journey towards the atmosphere. Hence, the Norwegian oil and gas sector is also contributing to the negative impact of fossil fuels due to its significant share of CO_2 emissions since the petroleum industry started in the early 1970s. In other words, Norway, as a net exporter of oil and gas, is contributing to the high atmospheric concentrations of CO_2. Although numbers for the first years of the petroleum activity in the North Sea are uncertain, it is nevertheless possible, based on available information, to affirm that emissions from the production and consumption of Norwegian oil amount to approximately 12 billion tons CO_2 (Ryggvik 2013: 17). Despite significant quantities of emissions, Norway has managed to inscribe its role as an oil and gas producer primarily as part of the international climate solutions. How the country has managed to reconcile its climate and petroleum policy discrepancies is largely the result of the consumption-based CO_2 accounting practices and market mechanisms on which the Kyoto Protocol rested. The Paris Agreement has kept the same approach. The next section, therefore, looks into the processes of framing climate change both nationally and internationally.

Framing the issue

I will draw upon Callon's conceptual pair of *framing* and *overflowing* (Callon 1998a), in combination with Latour's (1993) concept of hybrids, to analyze Norway's contradictions in the area of climate policies. Although the concepts seem to address different processes and, therefore, analytically serve different purposes, they actually have a lot in common. The idea of framing and overflow as two interdependent and performative movements is not so different from the realities that emerge through the work of hybridization. Framing occurs through demarcation when some objects, agents, and relations are taken into account, whereas others are separated or cut off, so to speak, from the issue at hand. Framing can, therefore, be understood as what specifically produces an issue as an issue, as it sets the boundaries for what belongs to the issue and what does not participate in its definition or enactment. In a certain way, we could think of framing as a way to make realities much more compact, limited, and, thereby, manageable, as it reduces the network of relationships: Some relations are taken into account, while others are ignored (Callon 1998b: 15). This brings us to the pairing process of overflow.

In economic theory, externalities are the outcome of market failures. According to Callon, this does not necessarily mean that the results were negative but, rather, "denote all the connections, relations and effects which agents do not take into account in their calculations when entering into a market transaction" (Callon 1998b: 16). In other words, the market did not operate or function as anticipated because some elements or situations were not accounted for. Callon, therefore, sees externalities as the result of an imperfect framing process (Callon 1998a). However, "airtight" framing is almost impossible due to the elevated costs of calculating for every possible eventuality and course of action. Accordingly, from the perspective of constructivist sociology, overflows are actually the rule, as framing always appears as an incomplete and fragile process (Callon 1998a: 252). The disentanglement through framing of an issue from its complete and extensive web of relations always contains the potential of overflows. Hence, the interactions, relations, knowledge, and actors defined and united by the frame can simultaneously belong to other frameworks and, thereby, circulate among a variety of locations. The framing process is also a way to settle a controversy (Callon 1998a), in this case, what constitutes the climate issue. However, controversies can always be reopened and thereby transit and settle within new frames. Therefore, in many cases, overflows can be quite productive. Conversely, the way issues are framed limits certain actions while simultaneously favoring other trajectories. To understand Norway's contradictory logics regarding its role as an oil and gas producer, we need to take a closer look at how climate change has been framed by national and international actors.

In the late 1980s, as a response to the World Commission on Environment and Development report, *Our Common Future* (1987), certain actors such as the Ministry of Finance and the central bank of Norway (Norges Bank), expressed concerns regarding the possibility of conflicting interests between specific goals of the report and Norwegian oil policies (Asdal 2011). In the following years, the Norwegian parliament voted in favor of significant emission reductions, besides implementing taxes on CO_2 from the offshore petroleum industry as policy measures in preparation for the upcoming Rio Conference (1992) (Ryggvik and Kristoffersen 2015). However, as the Earth Summit neared, it became increasingly clear to Norwegian authorities that a future international climate agreement could pose restrictions on the country's increasingly oil-fueled economic growth and thereby reduce its competitiveness (Asdal 2011). Hence, national climate policies were largely framed as an economic issue from the 1990s onward, which consequently had to be handled by economic authorities who emphasized the importance of seeking cost-efficient climate solutions (Asdal 2011; Hermansen and Kasa 2014). This implied promoting an international system that would allow Norway to fulfill its climate obligations abroad instead of having to perform costly emission cuts at home (Hermansen and Kasa 2014). Asdal maintains that three policy objectives underpinned Norway's position during the early climate negotiations: Emission targets should be established across borders so that countries could cooperate to reach them, the agreement had to include all GHGs, not only CO_2, and it had to enjoy wide support, including from major industrialized countries (Asdal 2011: 194). In other words, as things progressed on the international scene, Norway tried to influence the framing process by taking on an active role on several fronts. An important but complicated task for the Norwegian representatives was to position the idea of carbon offsets, which could be commercialized on an international market created for this purpose (Asdal 2011: 200). The Ministry of Finance started to enroll international actors by outlining guidelines and different policy proposals in order to present and promote the country's official position. Moreover, during several negotiation rounds before the Rio Conference, Norway tried to gain support from developing countries for its view (Asdal 2011: 201) to align their interests with those of the Norwegian government. If emission reductions could be obtained in collaboration across countries, buying emission quotas instead of performing expensive domestic reductions was clearly a cheaper alternative for Norway. Hence, national political and economic interests played a decisive role regarding how to design appropriate mechanisms that could simultaneously protect Norwegian industry, as well as the offshore oil and gas production, and enable the Norwegian government to fulfill its climate commitments. An important dilemma in this context was the impossibility of performing substantial cuts in the electricity sector since Norway was already using hydro-electric power (Reitan 1998 in Hermansen and Kasa 2014: 2). Without cheap alternatives at home, emissions trading and joint implementation became attractive policy solutions, as these mechanisms enabled Norway to keep the national petroleum economy disconnected from international climate challenges. Consequently, from an early stage of the international climate negotiations, it became crucial for several Norwegian actors to influence or have a say in the framing process. The transformation or alignment of interests to establish congruent visions of the world has to do with all interests being "temporarily stabilized outcomes of previous processes of enrollment" (Callon and Law 1982: 622). This implies that interests should be regarded more as stabilized effects of particular network relations than simply as background factors that explain the outcome. In short, interests work as ways to "impose order on a part of the social world" (Callon and Law 1982: 622). By framing climate change principally as an economic issue, enacted and shaped through market assumptions and economic theory, Norway was able to politically separate oil from climate change, despite several scientific reports working to link them together. This approach largely explains how the country has managed to maintain an internationally acknowledged green profile and simultaneously remain a significant oil and gas producer that shows no signs of preparing to phase out its dominant hydrocarbon industry. This situation can be observed in the following excerpt from white paper 28 (2010–2011), *An Industry for the Future – Norway's Petroleum Activities*, issued by the Ministry of Petroleum and Energy:

> The Government wants to combine Norway's role as a major energy producer with the ambition of being a world leader in environmental and climate policy, through continuing to exploit the petroleum resources while simultaneously pursuing efforts to streamline the activity on the continental shelf. The activity on our continental shelf should also be best-in-class when it comes to energy-efficient production of oil and gas.
>
> (2011: 7)

The logic seems to be that Norway's ambitions as an oil and gas producer do not conflict with pursuing a leading role in the international climate arena, as long as Norwegian oil is efficiently produced. The government and the petroleum sector itself often use the argument about Norwegian petroleum production being much cleaner than what is the case in other parts of the world as a way to inscribe the national oil industry as a legitimate part of climate solutions in a transitional phase. However, the difference between countries in CO_2 emissions from production[8] is rather insignificant if we take into consideration that the quantity of CO_2 emitted during combustion is approximately 40 times higher (Ryggvik and Kristoffersen 2015). It does not actually make any difference if the oil comes from the Middle East or Norway. This illustrates that the Norwegian argument is only made possible due to the current CO_2 accounting practices, as "the global ecosystem does not discriminate against the geographical origins of CO_2" (Bradshaw 2014: 17). A related aspect in this debate is that while the UNFCCC bases national inventories of GHG emissions only on emissions produced within sovereign territories, many developed countries benefit from "embodied-carbon" in commodities from different parts of the world, especially emerging economies in Asia (Davis and Caldeira 2010: 5687). Norway has, in other words, benefited from the current international CO_2 accounting practices, by which its oil and gas exports are calculated only as emissions in countries where the resources are consumed. However, those accounting practices have enabled the country to portray Norwegian oil as "clean" or as "low-carbon oil" when compared to petroleum resources extracted elsewhere. This has allowed the country to produce accounts about the Norwegian petroleum industry as having certain environmental advantages, thereby inserting it as an essential part of international climate solutions in national political discourse. Accordingly, it would be close to *unethical* to slow down the extraction rate as long as there are consistent need and demand. In line with this perspective a member of parliament from the Conservative Party argued:

I believe it is a completely pointless debate. If it was only a matter of closing down the petroleum production in Norway and that this would bring the green shift, then I think it would have been a very good debate, but that is not the case ... yesterday I went to listen to the director of IEA, the International Energy Agency, who said that the world needs it and I actually believe it is good environmental policy that this oil is extracted in Norway. We have the lowest CO_2 print on our extraction. As long as it is a product that the world demands, and that the world is going to buy, then I actually do not think it constitutes a problem.

Similarly, the current framing of climate change principally as a carbon emissions issue produces several possibilities of alignment and coordination of interests. Clean oil is not only a comparative advantage in a world that requires energy to fuel economic growth and development, but Norwegian gas can also replace coal, which is more carbon intensive and, thereby, directly make a contribution to the achievement of the European emission targets:

Norway is, and has always been, a stable and predictable supplier of oil and gas. In today's world, this is a competitive advantage. Gas can unite the European objectives of secure energy supply and reduced emissions of greenhouse gases. If gas replaces coal in production of electricity in Europe, this measure alone will suffice to fulfill the region's CO_2 objectives for 2020.
(Ministry of Petroleum and Energy 2011: 7)

If juxtaposed, the oil and gas arguments exhibit clear examples of *overflows* that are produced as direct effects of the framing process. The hybrid at work is the relationship between production and consumption of hydrocarbons that is differently used, depending on what is being performed: The "environmental advantages" of Norwegian oil or the positive contributions of Norwegian gas to address global climate change. When the first case is the center of the argument, stakeholders employ a production-based framing. In other words, what is important is what happens during extraction at a national level, and the consumption phase is kept outside the frame by highlighting that Norway is only responsible for emissions within its borders. As previously explained, this is made possible by the UNFCCC accounting practices. However, when the argument revolves around the importance of gas as part of "climate solutions", it is not a national, production-based framing that stakeholders hold on to; rather, they employ a consumption-based frame to do the work. Ryggvik and Kristoffersen summarize the Norwegian argument as follows: "When the issue is a massive increase in gas production, the argument is that it 'reduces' emission abroad. When it comes to production of oil, emissions from burned Norwegian oil abroad have not been taken into account at all" (2015: 266). Consequently, the double movement between global and national arenas, between production- or consumption-based frames, illustrates how economic considerations and, thereby, the oil and gas industry are the real issues at stake. Whereas the ongoing petroleum debate in Norway shows how scientific research is used to perform different political and economic scenarios, it also illustrates the versatility of scientific knowledge in political decision-making. Despite different climate reports stating that the large majority of fossil fuels must remain in the ground in order to reduce the dangerous concentrations of GHGs in the atmosphere, this information does not necessarily problematize the extraction and exports of fossil fuels. On the contrary, scientific knowledge about climate change can also constitute a useful political tool to redefine the role of the petroleum industry. Instead of being part of the problem, climate change is giving the national oil and gas industry a comparative advantage.

How oil became an endangered species

As we have seen, long before the signing of the UNFCCC Kyoto Protocol in 1997, the focus of climate change mitigation has been on emission reduction targets. In other words, the issue is framed as a carbon emissions problem; consequently, the emphasis is placed on how to effectively manage emissions (Princen et al. 2015). With this particular focus, several mechanisms and tools were developed within market logics: Emissions quotas, carbon permits, tradeable emission units, etc. The Protocol, therefore, largely inscribes climate change as emissions and carbon economics with very little explicit focus on fossil fuels as such. When framing an issue, a particular set of relations and connections are implicitly activated: If carbon emissions are the problem, reduced emissions and emissions management become a logical solution. In other words, *avoided emissions* as a possible political tool to address the climate issue stays out of the frame. This situation points towards a crucial aspect of framings, namely their ontological performativity, as they make certain trajectories and courses of action possible, while others become difficult or even impossible (Law and Singleton 2014). Fossil fuels play a significant role in the carbon problem, but they are, nevertheless, allowed to operate within configurations and relationships enacted outside the framework. Some authors like Princen et al. (2015: 7) have described these kinds of environmental policies as "end-of-pipe and cleanup approaches", as there is a series of phases from extraction to consumption that is left outside the relations that make up global climate change. This means that, from an economic viewpoint, "pollutants emerge after the goods are produced, as a by-product, an unfortunate yet unavoidable side effect" (Princen et al. 2015: 8). In short, economic theory regards them as externalities, that is, all those situations and connections that have not been accounted for.

Framing carbon emissions as the dominant problem, however, is largely a reductionist approach from an ANT perspective, as it cuts off and ignores a series of actors and entities that participate in originating the problem. Accordingly, trying to fix emissions only at the end of the chain when fossil fuels have already been combusted is a politically titanic and costly task, as new amounts of fossil fuels are extracted daily and start their conversion to become new emissions. The framework leaks in more than one way: First, the carbon market, embedded in economic thinking, did not perform as expected, as the relationship between supply and demand of carbon permits was far from optimal. The IEA has maintained that the inefficiency of the Kyoto Protocol was the result of a combination of factors: Few countries made commitments regarding the Protocol's second period, and too many project credits followed from the first period, which led to low carbon prices (IEA 2015a: 12). Second, by relying on market mechanisms and converting carbon into a tradeable commodity, the origin of the problem was made invisible. Pollution and emissions were performed as a *matter of economic transactions*, not as matters of concern, as market mechanisms produced a governable climate, at least in theory. Norway actively pursued a leading role early on in the international framing process, which in practice meant aligning or linking the future international agreement with Norway's position as a leading oil and gas producer. This was important for the Norwegian government since climate change policies had the potential of making petroleum assets increasingly problematic and unstable. From a Norwegian perspective, represented by the Ministry of Finance and other government agencies, international climate policies had the potential to endanger the country's oil wealth (Asdal 2014: 2119). In other words, *oil was becoming an endangered species*.

Another kind of framing was definitely possible, as the issue could have been framed with a more explicit focus on the role of fossil fuels or on climate and environmental justice. The indiscriminate extraction of fossil fuels as the main issue would have brought *non-extraction policies* to the forefront as an essential part of global climate solutions. While several scientific reports and carbon budgets emphasize that the majority of the world's fossil-fuel reserves constitute unburnable carbon, there is today very little debate internationally about what this actually means in practice. In other words, *which* reserves have to remain in the ground and *where* are not discussed as part of the UNFCCC framing. The Paris Agreement (2015) says nothing about these uncertainties; moreover, fossil fuels are not even mentioned in the document. This clearly indicates how climate change is framed within the international climate regime.

Norway's Kodak moment or the green shift

The informants' accounts portray the green shift as both a delayed and problematic process, but they also emphasize that Norway has all the necessary conditions to perform a transition. However, while private initiatives and market mechanisms are believed to be important tools in this process, state intervention is also required to properly orient the process and the transformation pace. Some informants emphasize that there is actually a lot to learn from the country's oil history when the state played an active role in directing the incipient industry by using political tools, legal frameworks, and a special taxation system. The accounts highlight the need to increase public expenditure in innovation projects and research on renewable energy alternatives. As for today, the oil and gas industry are compensated for their exploration investments with 78% in tax returns, based on the idea that it should be attractive for oil companies to explore on the NCS. The same taxation advantages do not apply to other sectors or industries. Consequently, several informants believe that an important strategy to accelerate the green shift is to take away the economic incentives that benefit the oil companies today and reorient them towards the development of green energy.

While Norway officially does not subsidize fossil fuels, several environmental organizations and non-governmental organizations (NGOs) emphasize that in practice, this is actually the case. When we talk about energy subsidies, we tend to think of developing countries and petroleum producers that forgo public revenues to finance part of the direct cost of fossil fuels to consumers. This kind of subsidy makes energy both accessible and affordable; however, it also distorts the market (Bradshaw 2014: 167). International organizations, such as the World Bank, IEA, and G20, tend to oppose direct energy subsidies because they tend to stimulate consumption and erode energy efficiency (Bradshaw 2014: 167). Conversely, it has been argued that this strategy provides access to low-income groups in developing countries as a means of reducing energy poverty and, thereby, boosts development. However, from a climate perspective, energy subsidies are highly problematic, as they motivate increased resource consumption, which means increased CO_2 emissions. The urgency to phase out fossil-fuel subsidies was addressed in a speech given by the former Norwegian Minister of Foreign Affairs, Børge Brende (2015), at COP21 in Paris:

> Facing a challenge of this magnitude, we must make use of *all* available tools to cut emissions and adapt to the changes that will come. What we should *not* do is undermine our own efforts. But fossil fuel subsidies are doing just that. They are negative climate finance. They contradict our common objectives of cutting emissions. They weaken our efforts to promote renewable alternatives. It is positive that we have committed to mobilizing USD 100 billion in climate financing by 2020. But it is a great paradox that fossil fuel subsidies amount to five times that figure. The International Energy Agency has estimated the value of fossil fuel subsidies worldwide at USD 490 billion in 2014. Developed countries are handing out fossil fuel subsidies with one hand and contributing

to climate financing with the other. Many developing countries are receiving climate finance while also subsidizing fossil fuels. This cannot continue.

This statement illustrates some of the contradictions involved in fossil-fuel subsidies and how the lack of coordination between international climate and energy policies results in outcomes that point in opposite directions. In this context, it is important to mention that while Norway does not specifically apply subsidies to end-consumption of fossil fuels, several existing mechanisms work as economic incentives applied to the oil industry. Environmental NGOs have, on several occasions, criticized the fact that oil companies are allowed tax deductions on all costs related to "exploration, research and development, financing, operations and decommissioning" (Norwegian Petroleum 2020), and they regard this as a way of subsidizing the Norwegian oil industry. By motivating continued exploration on the continental shelf and expanding the oil frontier in the Arctic, Norway is contributing to its own "carbon lock-in", which makes a green shift increasingly difficult to accomplish. Several informants believe that instead of pouring more resources into the petroleum industry, the government should take concrete action and give companies that are willing to develop and invest in alternative energy resources the same kind of economic incentives that have been used so far in the petroleum sector. Some accounts emphasize the need to directly subsidize renewables for a period as an active political tool to make green alternatives more competitive on the market as compared to fossil fuels.

Interestingly, the minister's speech portrays only fossil-fuel subsidies as problematic, not fossil fuels themselves. By enacting climate change finance, in competition, and thereby, in opposition to fossil-fuel subsidies, the *origin* of the problem is once more left out of the frame since keeping oil in the ground is not part of the political toolkit to cut emissions. The extraction of fossil fuels is *out-framed*, so to speak. By out-framing, I specifically refer to how national policies work to decouple fossil fuels and climate change: The relationship is evidently acknowledged scientifically, but politically it is only taken into account by making the relationship economically but not necessarily environmentally controllable. As opposed to externalities, which are all those situations and associations that have not been accounted for during the framing process, what I have chosen to call *out-frames* take certain entities and relations *highly* into account by specifically locating them as part of a different context. In other words, national climate policies produce a climate issue in which the petroleum industry is not involved, as it is specifically situated in a separate political framework. The alignment of international and national climate policies has made this *noncoherent* ordering a viable possibility. Law et al. (2013) have named noncoherent coordination as *modes of syncretism*. Parting from Latour's argument about modern purification and hybridization as parallel and coexisting processes, Law et al. explain that practices are always hybrid, or syncretic, which means they are never straightforward but tend to be rather fuzzy. In the case of the Norwegian oil and climate policies, a lot of work is performed to keep them apart since practices only appear as noncoherent when they actually meet. If this is not the case, syncretism remains only as a latent possibility (Law et al. 2013). By relying on different mechanisms such as temporal distribution and spatial segregation, noncoherent practices are kept apart (Law et al. 2013: 181). In short, one Norwegian minister announced the 23rd licensing round in January 2015; another minister signed the Paris Agreement in December the same year. However, on some occasions, despite a lot of effort to separate and distance the two logics in use, they can actually meet, and it is only in this encounter that "noncoherence becomes an issue" (Law et al. 2013: 180). This was the case when several Norwegian environmental organizations in the national media emphasized the contradiction between the government's commitment in Paris to a reduced carbon budget within a 1.5°C scenario and the licensing of new oil blocks in the Arctic (Barents Sea). A representative from the environmental organization Bellona described the syncretism of Norway's "doublethink" in the following way:

> The common message from the environmental movement was, okay, when we have established the 1.5°C goal, then this certainly must have some kind of consequence for the petroleum industry. We will have to suspend the 23rd concession round and stop licensing new oilfields. I believe the environmental movement managed to create a bit of buzz around it; well, okay, then this is certainly what it means. The end of the story was that Schjøtt-Pedersen, Erna Solberg and Tord Lien rapidly started a "fire extinction" and said no, this is not going to have any consequences for the Norwegian petroleum industry.[9]

A green shift without a significant transformation of the country's oil-fueled economy constitutes, according to several informants, a kind of *Kodak moment*. Similar to the Eastman Kodak Company that failed to foresee the importance of the digital camera and seize the moment, Norway's continuous high oil exploration and production level constitutes, according to several stakeholders, a missed opportunity to reorient and diversify the economy. Instead of investing heavily in renewable energy, in other words, what is believed to be the future, resources are allocated to an industry that necessarily has to be phased-out if climate goals are to be achieved. The recent opening of new oilfields will perpetuate the industry beyond 2050 when the country aspires to have become carbon neutral. By locking in labor, investments, research, and technological innovation in the petroleum sector, Norway could lose the opportunity to have a leading role in the development of renewable energy and corresponding technological solutions.

Notes

1   Norwegians often talk about the country's oil history as a "fairytale" implying that it has been an adventure leading to economic and technological success.
2   According to Latour, "black box" or "blackboxing" is a kind of simplification used by cyberneticians when "a piece of machinery or a set of commands is too complex. In

its place they draw a little box about which they need to know nothing but its input and output" (Latour 1987: 2-3)

3 All quotations from white paper 25 (1973–1974) are my own translations.

4 The former Norwegian Minister of Petroleum and Energy, Tord Lien, used this phrase when he described Norway during a lecture at Nord University on May 12, 2016. The topic was "Northern Energy – Developing Energy Resources in the High North".

5 The formal name of the Petroleum Fund is the Government Pension Fund Global. The investments are distributed in more than 9000 companies in 74 countries. Norges Bank Investment Management: https://www.nbim-no/en/.

6 The Norwegian title is *Petroleumsnæringens plass i det norske samfunn* (St. meld. no. 25, 1973–1974)

7 The IEA was founded as a direct response to the 1973–1974 oil embargo to help industrialized countries secure oil supplies from major disruptions.

8 Numbers from 2006 show that Norway has approximately 8 kilograms of CO_2 emissions per barrel of oil compared to 12 kilograms in the Middle East (Konkraft 2009).

9 Erna Solberg: Leader of the Conservative Party and Norway's Prime Minister since 2013.

   Karl Eirik Schjøtt-Pedersen: Director General of the Norwegian Oil and Gas Association from 2015 to 2020.

   Tord Lien: Minister of Petroleum and Energy from 2013 to 2016.

References

Asdal, K. (2011). *Politikkens Natur-Naturens Politikk*. Oslo: Universitetsforlaget.

Asdal, K. (2012). Contexts in action – and the future of the past in STS. *Science, Technology, & Human Values*, 37(4), 379–403.

Asdal, K. (2014). From climate issue to oil issue: Offices of public administration, versions of economics, and the ordinary technologies of politics. *Environment & Planning A: Economy and Space*, 46(9), 2110–2124.

Asdal, K. (2015). What is the issue? The transformative capacity of documents. *Distinktion: Scandinavian Journal of Social Theory*, 16(1), 74–90.

Bradshaw, M. (2014). *Global Energy Dilemmas*. Cambridge: Polity Press.

Brende, B. (2015, December 7). *Speech Given at COP21. Paris: Phasing out Fossil Fuel Subsidies*. Retrieved from https://www.regjeringen.no/no/aktuelt/fossil-fuels-subsidies/id24 66329/.

Callon, M. (1998a). An essay on framing and overflowing: Economic externalities revisited by sociology. In: M. Callon (ed), *The Laws of the Markets* (pp. 244–269). Oxford: Blackwell Publishers.

Callon, M. (1998b). Introduction: The embeddedness of economic markets in economics. In: M. Callon (ed), *The Laws of the Markets* (pp. 1–57). Oxford: Blackwell Publishers.

Callon, M., & Latour, B. (1981). Unscrewing the big Leviathan: How actors macrostructure reality and how sociologists help them to do so. In: K. Knorr-Cetina, & A. Cicourel (eds), *Advances in Social Theory and Methodology: Towards an Integration of Micro-And Macro-Sociologies* (pp. 277–303). London: Routledge & Kegan Paul.

Callon, M., & Law, J. (1982). On interests and their transformation: Enrolment and counter-enrolment. *Social Studies of Science*, 12(4), 615–625.

Carbon Tracker. (2011). *Unburnable Carbon: Are the World's Financial Markets Carrying a Carbon Bubble?* Retrieved from http://www-carbontracker.org/wp-content/uploads/2014/09/Unburnable-Carbon-Full-rev2-1.pdf.

Carbon Tracker & Grantham Research Institute on Climate Change and the Environment. (2013). *Unburnable Carbon 2013: Wasted Capital and Stranded Assets.* Retrieved from http://www.carbontracker.org/wp-content/uploads/2014/09/Unburnable-Carbon-2-Web-Version.pdf.

Chevalier, J.-M. (2009). Introduction. In: J.-M. Chevalier (ed). *The New Energy Crisis: Climate, Economics and Geopolitics* (pp. 1–5). Basingstoke: Palgrave Macmillan.

Davis, S. J., & Caldeira, K. (2010). Consumption-based accounting of CO_2 emissions. *Proceedings of the National Academy of Sciences,* 107(12), 5687–5692.

Eriksen, T. H. (2006). Innledning: Tryggheten og dens motstandere. In: T. H. Eriksen (ed), *Trygghet* (pp. 11–31). Oslo: Universitetsforlaget.

Finansdepartementet (1974). *Petroleumsnæringens plass i det norske samfunn (St. meld. No. 25, 1973–1974).* Retrieved from https://www.stortinget.no/no/Saker-og-publikasjoner/Stortingsforhandlinger/Lesevisning/?p=1973-74&paid=3&wid=c&psid=DIVL658.

Goldthau, A. (2011). Governing global energy: Existing approaches and discourses, *Current Opinion in Environmental Sustainability* 3(4), 213–217.

Hermansen, E. A., & Kasa, S. (2014). Climate policy constraints and NGO entrepreneurship: The story of Norway's leadership in REDD + financing. Working Paper 389, *CGD Climate and Forest Paper Series,* 15, 1–31.

IEA (2012). *World Energy Outlook.* Retrieved from https://www.iea.org/publications/freepublications/publication/WEO2012_free.pdf.

IEA (2015a). *Key Trends in CO_2 Emissions. Excerpt from: CO_2 Emissions from Fuel Combustion, 2015 Edition.* Retrieved from https://www.iea.org/publications/freepublications/publication/CO2EmissionsTrends.pdf.

IEA (2015b). *World Energy Outlook 2015, Executive Summary.* Retrieved from https://www.iea.org/Textbase/npsumWEO2015SUM.pdf.

IEA (2020). *World Energy Balances: Overview.* Retrieved from https://www.iea.org/reports/world-Energy-Balances-Overview.

IPCC (2007). *Climate Change 2007: Mitigation of Climate Change: Contribution of Working Group III to the Fourth Assessment Report of the Intergovernmental Panel on Climate Change.* Cambridge: Cambridge University Press.

IPCC (2014a). *Climate Change 2014: Mitigation of Climate Change: Working Group III Contribution to the Fifth Assessment Report of the Intergovernmental Panel on Climate Change.* New York: Cambridge University Press.

IPCC (2014b). *Climate Change 2014 Synthesis Report Summary for Policymakers.* Retrieved from http://www.ipcc.ch/pdf/assessment-report/ar5/syr/AR5_SYR_FINAL_SPM.pdf.

IPCC (2014c). *Summary for Policymakers. In: Climate Change. Contribution of Working Group III to the Fifth Assessment Report of the Intergovernmental Panel on Climate Change.* Retrieved from https://ipcc.ch/pdf/assessment-report/ar5/wg3/ipcc_wg3_ar5_summary-for-policymakers.pdf.

Kaarhus, R. (1999). *Conceiving Environmental Problems: A Comparative Study of Scientific Knowledge and Policy Discourses in Ecuador and Norway.* Oslo: NIBR.

Knorr Cetina, K. (2001). Objectual practice. In: T. R. Schatzki, K. Knorr Cetina, & E. Von Savigny (eds), *The Practice Turn in Contemporary Theory* (pp. 175–188). New York: Routledge.

Konkraft (2009). *Konkraft-rapport 5. Petroleumsnæringen og Klimaspørsmål.* Retrieved from http://konkraft.no/wp-content/uploads/2016/04/KonKraft-rapport-5-Petroleumsn%C3%A6ringen-og-klimasp%C3%B8rsm%C3%A5l-Lavoppl%C3%B8selig-siste-versjon.pdf.

Latour, B. (1987). *Science in Action: How to Follow Scientists and Engineers Through Society.* Cambridge, MA: Harvard University Press.

Latour, B. (1993). *We Have Never Been Modern*. Cambridge, MA: Harvard University Press.

Latour, B. (2000). When things strike back: A possible contribution of "science studies" to the social sciences. *British Journal of Sociology*, 51(1), 107–123.

Latour, B. (2005). *Reassembling the Social: An Introduction to Actor-Network Theory*. New York: Oxford University Press.

Law, J. (1994). *Organizing Modernity*. Oxford: Blackwell.

Law, J. (2004). *After Method: Mess in Social Science Research*. London: Routledge.

Law, J., & Singleton, V. (2014). ANT, multiplicity and policy. *Critical Policy Studies*, 8(4), 379–396.

Law, J., Afdal, G., Asdal, K., Lin, W.-y., Moser, I., & Singleton, V. (2013). Modes of syncretism: Notes on noncoherence. *Common Knowledge*, 20(1), 172–192.

Ministry of Climate and Environment (2014, December 10). *Green Shift: Climate and Environmentally Friendly Restructuring*. Retrieved from https://www.regjeringen.no/en/topics/climate-and-environment/climate/innsiktsartikler-klima/green-shift/id20 76832/.

Mol, A. (2002). *The Body Multiple: Ontology in Medical Practice*. Durham: Duke University Press.

Norges Bank Investment Management (n.d.). Retrieved from https://www.nbim.no/en/.

Norli, K. (2016, April 19). Petroleumsfagene er årets taper i søknadstallene fra samordna opptak. *E24*. Retrieved from https://e24.no/privatoekonomi/i/ddW0Rj/petroleumsf agene-er-aarets-taper-i-soeknadstallene-fra-samordna-opptak

North, D. C. (1990). *Institutions, Institutional Change and Economic Performance*. New York: Cambridge University Press.

Norwegian Ministry of Petroleum and Energy (2011). *Meld. St. 28 (2010-2011) Report to the Storting (White Paper). An Industry for the Future-Norway's Petroleum Activities*. Retrieved from https://www.regjeringen.no/globalassets/upload/oed/petroleumsmeldingen_2011/oversettelse/2011-06_white-paper-on-petro-activities.pdf.

Norwegian Petroleum Directorate (2010, December 15). *10 Commanding Achievements*. Retrieved from https://s3.amazonaws.com/rgi-documents/e3cbbfde7c90c60753b47 7e84627ee06dd50ae25.pdf.

Norwegian Petroleum Directorate (2015, January 20). *23rd Licensing Round: Announcement*. Retrieved from http://www.npd.no/en/Topics/Production-licenses/Theme-articles/Licensing-round/23rd-Licensing-round/Announcement/.

Norwegian Petroleum (Updated 2020, December 5). *The Petroleum Tax System*. Retrieved from http://www.norskpetroleum.no/en/economy/petroleum-tax/.

Princen, T., Manno, J. P., & Martin, P. L. (2015). The problem. In: T. Princen, J. P. Manno, & P. L. Martin (eds), *Ending the Fossil Fuel Era* (pp. 3–36). Cambridge, MA: MIT Press.

Ryggvik, H. (2013). *Norsk Olje og Klima: En Skisse Til Nedkjøling*. Oslo: Gyldendal Arbeidsliv.

Ryggvik, H. (2015). A short history of the Norwegian oil industry: From protected national champions to internationally competitive multinationals. *Business History Review*, 89(1), 3–41.

Ryggvik, H., & Kristoffersen, B. (2015). Heating up and cooling down the petrostate: The Norwegian experience. In: T. Princen, J. P. Manno, & P. L. Martin (eds), *Ending the Fossil Fuel Era* (pp. 249–275). Cambridge MA: MIT Press.

Verbruggen, A., & Van de Graaf, T. (2015). The geopolitics of oil in a carbon-constrained world. *International Association for Energy Economics*, 2, 21–24.

3   Translating Lofoten

After a brief introduction, this chapter sets out to analyze the ongoing sociopolitical process and environmental controversy attached to the possibility of opening the Lofoten area to oil drilling and, thereby, expanding the oil frontier to territories classified as vulnerable in several research documents and the integrated management plan. An important point of departure is, therefore, tracing some of the texts and knowledge systems that govern the Lofoten area. As part of this analysis, I will specifically look into how tensions between science and politics and oil and environment are enacted and translated. Finally, the last part of this chapter is dedicated to analyzing Lofoten as a heterogeneous territory and how this situation produces conflicting and overlapping network translations.

Since the beginning of the 1970s, the expansion of the oil frontier towards the north has been a recurrent theme in the Norwegian political debate. Today, the controversy revolves around the possible consequences of petroleum activity in highly sensitive and valuable areas such as Lofoten, Vesterålen, and Senja (LoVeSe) and the Arctic. As part of this debate, we find the environmental movement claiming that the government has politically "moved" the ice edge[1] to facilitate further expansion of the oil industry and, thereby, opened up new blocks for licensing in unexplored areas in the Barents Sea. However, during the yearly Arctic Frontiers[2] conference in Tromsø in January 2015, Prime Minister Solberg argued that a new definition of the ice edge is not politically motivated but the result of updated scientific data, which indicates that the ice is pulling back. In other words, "the ice edge has moved itself", according to Solberg (Stavanger Aftenblad 2015). Environmentalists and some political parties, mainly minority parties that enabled the formation of the current government due to their parliamentary support, consider this position inconsistent and irresponsible as they emphasize that the sea ice is melting due to global warming that is largely caused by the burning of fossil fuels. In other words, further and intensified extraction in new frontier areas is only going to aggravate the problem and is, therefore, negative for global climate change mitigation. Regarding this contradictory situation, a representative from Friends of the Earth Norway made the following comment during an interview in 2015:

> Well, you can say that there is nothing particular about the oil outside Lofoten, Vesterålen, and Senja compared to other oil, which is correct, but

today we know that as much as four-fifths of already discovered fossil energy resources must remain in the ground if we are going to avoid the most serious climate change consequences. We, therefore, believe that we have to manage to leave the oil in our most vulnerable sea areas. The Executive Secretary of the UNFCCC [United Nations Framework Convention on Climate Change] Christiana Figueres was recently in Norway. She said that if the world is going to manage to limit the rise in global temperature to maximum 2°C, the world has no room for oil from the Arctic.

The statement addresses two of the most important arguments in the ongoing national and international climate debate: First, all known fossil-fuel reserves cannot be extracted if there is going to be any possibility of not surpassing a 2°C rise in temperature (the intention of 1.5°C is an even bigger challenge). Second, since we necessarily must keep most petroleum reserves in the ground, a good start would be hydrocarbons located in environmentally vulnerable areas such as LoVeSe and the Arctic. According to the environmental movement, it is counterintuitive to open up these locations for the petroleum industry if the world already faces a situation in which the large majority of fossil-fuel reserves constitute unburnable carbon. The controversy is, therefore, largely concerned with the uncertainties related to complex ecosystems and the use of particular territories. In other words, the climate change issue is frequently performed by the informants' accounts as a matter of how territories are classified and employed in contradictory ways. Consequently, one of the key elements in this debate is how policies coordinate, assemble, and negotiate understandings of specific territories with both political and environmental purposes, a situation that often generates tension and *noncoherence* (Law et al. 2013) between different policy goals.

In the past, however, the territorial controversy was linked to the challenges of petroleum activity above latitude 62. During the first years of oil production in the North Sea, several white papers produced by the Ministry of Industry presented the crossing of latitude 62 more as a technical problem than a political one (Rommetvedt 2014: 49). In fact, disagreement existed among different actors about when to start exploration further north, as some believed that the Norwegian companies needed time to acquire the necessary competence to carry out the task (Rommetvedt 2014: 47). Overall, moving north was principally regarded as a possibility to address regional policy needs in northern Norway, which was stated early on in the *oil commandments*: "A pattern of activities must be selected north of the 62nd parallel, which reflects the special socio-political conditions prevailing in that part of the country" (Norwegian Petroleum Directorate 2010: 4).

The focus changed with white paper 25 from 1974, issued by the Ministry of Finance, as it analyzed several societal and environmental implications that had not been included in previous white papers. This wider scope has to do with the Ministry of Finance largely working as the government's coordinating ministry, a position that necessarily requires a broader approach than policies generated by a single sector (Rommetvedt 2014: 50). Accordingly, the white paper was produced in collaboration with several other ministries, which added participants and related fields to the process. This situation evidently made the political decision-making much more complex, as the white paper presented not only economic opportunities and uncertainties but also environmental risks related to offshore petroleum production. With white paper 25, environmental and territorial considerations specifically became part of the political agenda, as potentially conflicting interests between the oil industry and the fisheries were taken into account. While the first white papers enacted latitude 62 principally as a technical frontier, in other words, as a practical matter, white paper 25 emphasizes the multiple implications of oil in a variety of fields. With this broader problematization, latitude 62 also became a political and environmental boundary: In 1976, environmentalists, local opponents, and fishermen's associations joined forces with some political parties and founded Oil Action in the North (Larsen and Aarsæther 1977; Rommetvedt 1991 in Sande 2013: 69). Nevertheless, despite uncertainties regarding environmental risk, safety measures, and possible territorial conflicts between existing industries and the petroleum sector, several political parties supported the decision to move north of latitude 62. This situation changed in 1977 with the blowout at the Ekofisk oilfield (the Bravo platform), as the incident produced a general agreement on the need to postpone the intention to move northwards. However, the blowout did not have the long-term political effect that environmentalists hoped for, since the incident occurred in open sea far from the shore and left no visible trace of the oil spills after a few weeks, despite the severity of the accident (Ryggvik and Kristoffersen 2015: 255). Although safety and environmental challenges had increasingly become part of the petroleum debate, especially after 123 oil workers lost their lives at sea when the Alexander Kielland platform capsized in March 1980, a proposal to further postpone oil activity north of latitude 62 lost by a majority vote in parliament. Hence, the Treasure Seeker oilrig started drilling on Tromsøflaket bank in July 1980 (Rommetvedt 2014: 57).

The inscription of Lofoten in policy and research documents

The existing knowledge base for the Lofoten area is rather extensive, as a considerable number of research reports and policy documents have been produced over the years since the government (the Bondevik II Government) announced in 2002 the decision to create a comprehensive management plan for the Barents Sea–Lofoten area. Accordingly, the plan was ready in 2006 and thereby became Norway's first management plan for a marine environment.[3] The plan is part of a continuing process with programmed revisions based on an expanding and evolving knowledge base. Updated versions came in 2011, 2015, and the latest revision of the *Integrated Management Plan for the Barents Sea–Lofoten Area* was ready in April 2020.[4]

Hence, the 2002 white paper, *Protecting the Riches of the Seas*, which set the management plan process in motion, came about partially as a response to international requirements to develop ecosystem-based management plans for ocean areas with a much more holistic focus on the ecosystems (Knol 2010c;

Olsen et al. 2016). An equally important driver was the oil and gas industry's aspiration to access new areas for exploration and extraction activities in the north (Knol 2010c; Olsen et al. 2016). Consequently, the integrated management process addresses one of the main political concerns tied to the territorial expansion of the petroleum industry that dates back to the early 1970s, namely, how to achieve a balance between sectors and industries that operate or wish to operate within the same territory, as stated in the 2002 white paper:

> The oil industry is moving closer to shore and more vulnerable areas. Shipping along the Norwegian coast is on the increase, thus increasing the risk of accidents. We now also know more about the vulnerability of our marine and coastal environment. All this means that conflicts between different user interests will increase in the years to come.
>
> (Ministry of the Environment 2002: 16)

While circumstances such as international trends oriented towards integrated ecosystem-based ocean management and a petroleum sector pushing for access to closed territories can be regarded as important backdrops for the development of the plan, the process itself largely produced its own contextual setting. By linking and drawing circumstances together as the previous excerpt shows, these elements simultaneously became context and part of the plan's inner workings. This means that we should not simply regard the document as a direct outcome of these international and national circumstances. Science and technology studies (STS) have a long tradition of problematizing a very common practice in the social sciences: that is, to explain a situation or event as a direct consequence of something that is believed to exist beyond or outside the event itself. A common assumption is that we contextualize an issue or situation by placing it in what is believed to be its broader context. Regarding this situation, Latour (2005), in a rather ironic fashion, describes social scientists' tendency to move away from local interactions and seek their explanation in exterior structures or events as some sort of *salto mortale* from the issue towards its contextual background. Instead, we are invited to focus on the way "contexts are gathered, summed up, and staged *inside* specific rooms into coherent panoramas *adding* their many contradictory structuring effects to the sites to be 'contextualized' and 'structured'" (Latour 2005: 191). Consequently, another possibility is to think of contexts as a relational dimension or, more specifically, as a set of relations, which means that a situation connects or relates to other situations or events and, thereby, extends the network. In short, certain relations are activated when contexts are evoked.

The Barents Sea–Lofoten area management plan brings together and works on several conflicting contexts simultaneously. International trends regarding integrated ecosystem-based management and pressure from the oil industry are only two of the contexts that operate in the plan. Another significant driver is global climate change and its cumulative effect on the ocean's ecosystems. However, with a large number of directorates, research institutes, organizations, and other stakeholders involved in the process, we can only imagine the number of contexts that are being worked upon and translated in the text. Since the management plan is not a fixed document but one that travels with time as updates and revisions are programmed periodically, based on an expansion of the knowledge base, new contexts will probably be combined by the document together with existing ones in the future.

The Deepwater Horizon accident in the Gulf of Mexico is an interesting example of how a context both delayed and was woven into the updated version of the management plan. While the plan's revision had to be finished at the end of 2010, the critical incident and oil spills in the Gulf of Mexico that same year made the government postpone the process so that new knowledge could be assessed and included in the updated plan, which was ready in 2011 (Hoel and Olsen 2012: 90). The Deepwater Horizon incident "acted as a caution to the development of the oil industry and was central in the political debate leading to the new zoning plan for the area" (Hoel and Olsen 2012: 91). This situation demonstrates how contexts are not necessarily an exterior reality that surrounds an issue but directly become contexts as they are combined, worked upon, and taken into account. According to Asdal and Moser (2012), the term "contextualizing" does not, however, properly convey the way contexts are combined and overlap and, thereby, are "worked into" and transformed by the issue at hand. Alternatively, they propose the term "*contexting*" when referring to how "contexts are being made together with the objects, texts, and issues at stake" (Asdal and Moser 2012: 303). Drawing on the idea of contexting can help us understand how the management plan, by integrating several contexts, enacts conflicting political and environmental realities that it simultaneously pursues to unite in its overarching goal:

> The purpose of this management plan is to provide a framework for the sustainable use of natural resources and goods derived from the Barents Sea–Lofoten area and at the same time maintain the structure, functioning, productivity and diversity of the area's ecosystems.
>
> (Ministry of the Environment 2011: 5)

Although the management plan seeks to reconcile both protection and utilization of the environment by emphasizing the goal of economic exploitation of the natural resources through sustainable use, this also constitutes the intersection where conflicting political positions play out both nationally and locally. With the dual intention of "facilitating value creation and maintaining the high environmental value of the area" (Ministry of the Environment 2011: 5), the Barents Sea–Lofoten plan is an attempt to balance economic and environmental interests. In other words, it tries to lay the groundwork for harmonious coexistence between diverse economic activities, such as the fisheries, oil and gas industry, maritime transport, local tourism, and the environment.

Interestingly, the management plan's dual purpose provides those who oppose opening the LoVeSe area to petroleum production, as well as those actors who favor oil drilling, with important lines of argument. The temporary oil moratorium in these areas means that for both sides in the controversy, every new parliamentary period constitutes new political opportunities and uncertainties. The following statement by a representative from the Norwegian Fishermen's Association illustrates how conflicting positions between the oil industry and the traditional fisheries are perceived:

> I don't know if it's actually due to short-term thinking because the politicians who deal with these things also think [of] economy in a long-term perspective and they probably have an idea about how this will contribute positively to the Norwegian welfare state. On the other side, they forget that it can have rather negative consequences for the fisheries and that we could experience a drop in fish stocks as well as in the number of fishermen and the general value creation around the activity. It is clear that at least Tord Lien,[5] as the oil industry's main representative, downplays the fisheries because he wishes to start something outside Lofoten and Vesterålen.

Whereas many stakeholders believe coexistence in the waters outside Lofoten and Vesterålen is impossible, based on the perceived risk to the fisheries and, thereby, their economic activity, others are positive about opening the area and emphasize the possibility of employment, value creation, and positive spin-off effects for the region. However, most informants consider the possibility that territorial conflict will be the decisive factor, which means that the territory's practical organization is what ultimately will decide the outcome of the controversy. This position can be observed in the following argument provided by a member of parliament from the Conservative Party:

> I believe the territorial conflict will be the decisive factor in the end. If we imagine that we get an impact assessment … then I think it will be a conflict over territory, the good territories in those areas where fishing and catch are to take place. Evidently, if we do not manage to combine them, then I think the fisheries will win. I think it will be the territorial conflict at the seafloor that will decide if there will be [oil] activity developed or not. Down at that level, actually. Is there any possibility of combining, or how much territory will the petroleum activity occupy of those areas that are the richest fishing fields in Norway?

A major goal behind the integrated management plan was, hence, to establish consensus between different actors and stakeholders by developing a set of conditions and tools that could enable coexistence in the area. A key strategy during the plan's development was, therefore, to involve a large number of entities and actors with a socioeconomic, political, or environmental stake in the issue.

In fact, before the integrated management plan for the Barents Sea–Lofoten area was passed in parliament, the governance of this sea area had been characterized by a system of sectorial administration, which led to a certain level of policy and management fragmentation. Consequently, applying a new cross-sectorial approach made it necessary to coordinate and assemble knowledge that existed spread among different sectors and research institutes. The short timeframe (2002–2006) made it difficult to produce updated knowledge before the government implemented the plan. During the process, however, the working groups identified several knowledge gaps and priorities (Olsen et al. 2016) to strengthen the knowledge base for the plan's revision in 2010.

Several uncertainties related to identified knowledge gaps thus landed the government on the decision to not open the Lofoten area to petroleum activity in 2006 (Olsen et al. 2016: 298). Three fields were given priority for the programmed update of the plan: (1) mapping of the seabed through the MAREANO program, (2) survey and monitoring of seabird populations through the SEAPOP program, and (3) seismic data on petroleum reservoirs with a focus on LoVeSe and *Eggakanten* (the edge of the continental shelf). The document states that these areas were selected "because they were of interest for the oil and gas industry and had also been identified as particularly valuable and vulnerable" (Ministry of the Environment 2011: 6).

Moreover, the inscription of Lofoten as a highly valuable and vulnerable area has become an important argument for political bargaining embraced by political parties and civil society groups that oppose opening the region for oil drilling. In other words, this particular status makes scientific knowledge and, consequently, the Barents Sea–Lofoten management plan important *allies* that opponents point to when arguing the potential risks of opening up the territory. Stakeholders continuously negotiate and renegotiate Lofoten's status as a vulnerable and valuable area and, thereby, negotiate the implications of these categories. Those who welcome oil drilling as positive for the region emphasize that environmental risk is something Lofoten already deals with but in the absence of proper safety measures. Therefore, the entrance of the oil industry could actually protect the region by making it less exposed to incidents and oil spills. A local politician from the Conservative Party emphasized this argument:

> If you define Lofoten as vulnerable and it is a valuable territory … if this is the point of departure, we can minimize the risk we have today because by opening up for oil activity, there will be some, what should I call it, frameworks that make it possible to establish environmental monitoring. We need readiness measures etc., which we currently do not have in Lofoten, but the risk is there since we have huge tank ships full of oil, huge ships carrying iron ore passing Lofoten as they round Lofotodden and enter Narvik that is a big ore port. We have big cruise ships coming close to our shores and they have heavy crude.

The People's Action for an Oil-Free Lofoten, Vesterålen, and Senja[6] is an example of a stakeholder that specifically highlights the significance of the knowledge contained in the management plan as it takes nature's carrying capacity into account. Accordingly, the role of scientific advice in the petroleum controversy regarding Lofoten is a decisive factor in the debate, as the following comment provided by a representative clearly shows:

The combination of important nature values is the reason this territory is classified as vulnerable and valuable in these management plans, that is, in the general plans regarding how we are going to manage the sea areas. The institutes I have talked about have reached these conclusions. We have environmental research saying that it is not a good idea and then there is the need to protect the interests of the fisheries; well, we have to protect both the biology and the economy in all this.

Even if today's science and the management plan largely comprise important political tools to reject the pressure from the oil industry to gain access to closed territories, there are informants who emphasize that uncertainty surrounds future revisions of the management plan. Whereas the current plan classifies the Lofoten area as a particularly valuable and vulnerable territory, this situation could change in future plan updates. As an evolving document that moves along with an expanding knowledge base, Lofoten's status could potentially change as new knowledge gaps are continuously being produced. Interestingly, the management plan process demonstrates that gaps in existing knowledge are not simply a matter of identification or discovery but are the outcome of negotiations and priorities that largely frame the process. Accordingly, the construction of particular knowledge gaps that research must fill in is one of the main crossroads where science and politics intersect and meet (Knol 2010a). In fact, an important purpose during the whole management plan process was to produce knowledge that could inform political decision-making (Knol 2010b: 246). Another significant aspect is that the inscription of knowledge gaps provides the document with agency, as it coordinates and distributes responsibilities between entities, actors, programs, and research institutes that will carry out new knowledge production. Since scientific uncertainties call for a precautionary approach towards economic activity in this sea area, the document performs the territory as overlapping translations in an attempt to integrate several fields of conflicting interests.

Following the perspective of Law et al. (2013), the political process surrounding the development of the integrated management plan, as well as the plan itself, can be regarded as an example of *syncretic modes of ordering*. The next section is, therefore, dedicated to the analysis of the Barents Sea–Lofoten management plan as an attempt to *domesticate* the resource conflict in the north, where concerns regarding the environment and competing industries meet. Production of scientific knowledge in this process is an important political tool that enables both negotiation and coordination of conflicting positions and logics.

The assemblage of politics and science

In 2011, the government upheld its 2006 decision not to open the Lofoten area nor carry out an impact assessment under the Petroleum Act. However, as part of a political compromise due to parliamentary coalitions, the updated management plan announced the need to further expand the knowledge base. In other words, it initiated a new process of knowledge gathering. The emphasis was placed on aspects related to economic activities: The Ministry of Petroleum and Energy would be in charge of researching the potential impacts of oil activity in the closed areas of Nordland IV, V, VI, VII, and Troms II. Similarly, the Ministry of Trade and Industry, the Ministry of Fisheries and Coastal Affairs, the Ministry of the Environment, and the Ministry of Local Government and Regional Development would research spin-off effects of increased economic activity in sectors such as the fishing industry and tourism.[7] While this information would be included in the next revision of the management plan, the document also states that it could be useful if an impact assessment for petroleum activities is carried out in the future (Ministry of the Environment 2011: 135, 137). Establishing and prioritizing gaps in knowledge is far from being an "innocent" scientific procedure. On the contrary, the Barents Sea–Lofoten area is an illustrative example of how controversial political issues are accompanied by scientific research in an attempt to settle and anchor decisions, the organization, and use of the territory in this particular case.

As several political, environmental, and socioeconomic uncertainties are attached to oil and gas production in the north, knowledge becomes a practical mechanism by which politics seeks robustness and solidity. While science informs the management plan, and its consensus-based approach is meant to unify and overcome discrepancies, there is always ambiguity. In Latour's words, "any laboratory observer knows, most of the phenomena depend upon which measure to read, or which to believe in the case of discrepancy" (1986: 23). The tension between exploitation/use on the one hand and conservation on the other emerges, as I have already discussed, in the encounter between science and political decision-making. Regarding these matters, Knol emphasizes the importance of exploring "how integrative knowledge is produced", despite a series of uncertainties and stakes and how knowledge is "incorporated and *translated* into a policy context" (2010c: 253).

Although the management plan process is very much about translating scientific research into policies, the translation process has a bidirectional character, as politics and policy requirements are also being translated into scientific research, which opens up new gaps in existing knowledge. As we have seen, these gaps in knowledge are highly political. As scientific research is meant to settle, or at least appease the political controversy related to possible petroleum activity in environmentally sensitive areas, uncertainties are followed up with more knowledge production. Whereas knowledge gathering has become a recurrent mechanism to balance conflicting perspectives, the final decision about the future of unopened territories, such as the Lofoten Islands, is not only a matter of scientific facts but also a question of values linked to environmental practices and ideas about what constitutes a political good. After more than a decade of researching the area, many actors argue that the requested knowledge is already available. Regarding this situation a senior advisor from the Norwegian Coastal Administration stated:

> There is enough [information] on social economy, on consequences for tourism, consequences for the fisheries, on aspects of a more technical nature, on

nature values, on identity and culture, on demographic development. You will find everything. I don't think there are almost any issues left to discuss. So, the reason NHO (the Confederation of Norwegian Enterprise), for example, says that we need an impact assessment is because it's a requirement to open up.

Nevertheless, politicians continue to demand more knowledge instead of analyzing the trade-offs between value creation and protecting valuable nature (Olsen et al. 2016: 299). The oil lobby often claims that an impact assessment under the Petroleum Act is a necessary step in order to get *the facts straight*. Conversely, several stakeholders question whether it is really a matter of obtaining more knowledge. The statements below highlight the hybrid character of these knowledge claims:

> The work done regarding the Management Plan for the Barents Sea is actually a great job. This means that we have a very good knowledge base, so an impact assessment will be additional to all the other things. So, it is very much a question of which values you fight for in society.
> Well, we are not really there yet, since we have not conducted a real process on these things. I think that, of course, based on ideological reasons or based on the knowledge each and every one has, we can say yes or no. But I think it is very much about ideology.

The informants' comments emphasize that an impact assessment or knowledge gathering, in general, is not necessarily a neutral process, as it also has a strong political drive. This means that knowledge practices, as part of our world-making efforts, are also a matter of ideology and values. This perspective resonates with Young's (1992) view on the science/ideology divide, namely, that ideology is an essential part of scientific practice. However, it is important to highlight that this does not necessarily mean that science directly produces biased knowledge due to scientists' political or ideological positions. Alternatively, ideology can be seen, and this is Young's view, as a constitutive element of knowledge production, since particular ways of conceiving nature influence the ways natures are produced.

Despite modern purification efforts (Latour 1993), science and politics go hand in hand when producing our worlds. In the case of the Barents Sea–Lofoten Area Management Plan, it produces a sea territory that is both a nature object and a political object, as the very process of producing demarcated, calculated, mapped, and monitored nature requires political action and policy measures. In other words, nature objects are political from the outset. We can, therefore, think of nature objects, or objectified nature, as the outcome of political negotiations and decision-making processes: Objectifying nature through different scientific procedures and practices constitutes the basis of environmental governance. This is how nature becomes manageable in different policy areas.

Moreover, the management plan states that the delimitation of the ocean area was the result of combining scientific knowledge about the marine ecosystems on the one hand and administrative considerations on the other. The area included in the plan is a vast territory that covers 1,400,000 km^2, which is almost four times Norway's land-based area (Ministry of the Environment 2006: 17). The territory extends from the Lofoten Islands in the south to the Arctic Ocean in the north and from the Norwegian Sea in the southwest to the delimitation line between Norway and Russia in the Barents Sea to the east (see Figure 3.1). Furthermore, the document explains that the Lofoten Islands were included as part of the management plan territory because of "the close ecological relationship between fish stocks here and in the Barents Sea" (Ministry of the Environment 2006: 17). One of the most relevant species is the Northeast Arctic cod that migrates towards Lofoten and Vesterålen, where spawning occurs every year from February to April. From these waters, eggs and larvae drift along the coast, ending up in the Barents Sea. The cod stock is, therefore, an organizing and connecting entity that participated in establishing the southern boundaries of the management plan territory by linking the Lofoten area to the Barents Sea through its spawning and migration patterns.

The other factor that contributed to defining the territory had more of an administrative character, as I have mentioned already. While ecosystems and lifecycles naturally operate on both sides of the Norwegian-Russian delimitation line, the plan area only covers the part of the Barents Sea where Norway has jurisdiction. After decades of controversy regarding the exact delimitation, Norway and Russia finally came to an agreement in 2010. However, increased pressure from commercial activity (shipping, the fishing industry, and oil and gas activity), together with challenges such as climate change and ocean acidification, may increment the cumulative effects on the sea environment (Ministry of the Environment 2011: 142). Consequently, there is a growing need to coordinate Norwegian and Russian policies in this field. The two countries are, therefore, collaborating on the development of a shared monitoring program for the whole Barents Sea (Ministry of the Environment 2011: 142). Furthermore, the end of the long-running demarcation dispute is also significant from a geopolitical perspective, as clarity on marine territorial jurisdiction also means predictability regarding economic development and resource exploration and extraction in this sea area.

It is important to remember in this context that Norway's oil production peaked in 2000, while by 2008, it had fallen by more than 32% (Ryggvik 2009: 145). Due to several maturing oilfields[8] on the continental shelf, expanding exploration into new areas became increasingly important for the Norwegian government. The agreement on the division line opened up new possibilities of cooperation regarding the exploitation of petroleum resources that could be located across the division line (Ministry of the Environment 2011: 13). The Arctic is, therefore, geopolitically significant as it implies co-presence with Russia. Conversely, the Arctic has increasingly become a "central part of Norway's strategy to retain its role as a major, secure exporter of hydrocarbons for several decades to come" (Emmerson 2013: 213). In other words, the Arctic (and the LoVeSe area for that

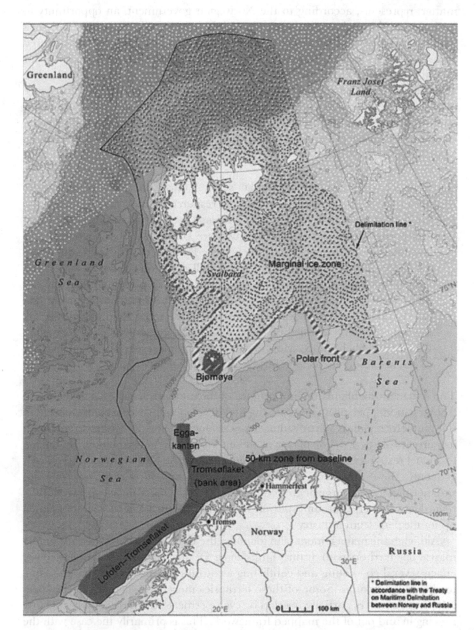

Figure 3.1 Particularly valuable and vulnerable areas in the Barents Sea–Lofoten
area. Map data: Norwegian Polar Institute, 2011. Source: Ministry of the
Environment, 2011.

matter) represents, according to the Norwegian government, an opportunity to prolong the Norwegian petroleum age despite oilfields on the continental shelf with declining production. For the 23rd licensing round in 2015, new unexplored areas located in the southeastern part of the Barents Sea were opened up for the oil companies. This was the first time in 20 years that the government gave access to new frontier areas for exploration. The strong focus on the importance of accessing new areas in the north in order to maintain Norway's role as a major energy producer can be observed in the following excerpt from a lecture given by the former Minister of Petroleum and Energy, Tord Lien, at Nord University in May 2016:

> If we are to continue being a huge exporter of oil and gas in [the] decades ahead, we are totally dependent on the resources outside northern Norway, not only in the Barents Sea but also acreage around, for instance, the Aasta Hansteen. Last year the gas infrastructure connecting the Norwegian gas resources to our European customers passed the polar circle. Did you know? The Aasta Hansteen field development will come on stream in 2018. The gas pipeline moving the gas to Europe is already in place and, thereby, we have connected our Arctic gas resources directly to the European market … My point being that we have produced little more than half of the oil, but approximately a third of the gas resources present on the Norwegian Continental Shelf (NCS). Most of what is remaining, especially when it comes to gas, are resources located outside northern Norway.

The statement highlights the importance of the oil and gas deposits in the north from a national perspective: Accessing and developing hydrocarbon resources in the northern region is portrayed as a key strategy if Norway wishes to maintain its current role as a major player in the international energy market. By linking fossil-fuel resources in the north to national exports and, thereby, to the macro-economy, these areas are enacted as part of a *national economic territory*. This situation generates a large degree of uncertainty as to what can be expected locally regarding value creation and spin-off effects if areas such as LoVeSe are opened up for the petroleum industry.

Although the management plan identifies the Barents Sea–Lofoten area on the map as one territory with distinct limits and a specific extension, it simultaneously enacts several co-existing and conflicting territories that operate within the same territorial boundaries. Some of these territories more or less coincide with each other, while others partially overlap or even permeate the territory's borders by moving in and out of the mapped framework. This is primarily the case with the management plan's ecological territory that due to complex interactions, lifecycles, and seasonal variability exists and lives on both sides of the mapped frontiers. While political maps work to perform settled or fixed dimensions of space, the ecosystems do not work by the same modus operandi, as their temporal and spatial strategies are completely different. Evidently, for the purpose of political decision-making and environmental governance, the complexities of more "fluid" territories are often difficult to handle, as they are too dispersed or distributed to be included in a single management plan or framework. The solution, therefore, is to cut off relations and arrange them in different management and policy documents. The following excerpt from the management plan illustrates how a combination of administrative and scientific logics underpin the management plan's spatial organization:

> Activities in the coastal zone on the landward side of the baseline that do not affect the sea areas outside the baseline have not been included, as coastal zone management involves problems of a different nature and to discuss these here would not serve the purpose of this management plan. However, impacts on the coastal zone caused by activities in the Barents Sea–Lofoten area, for example acute oil pollution, have been included.
>
> (Ministry of the Environment 2006: 17)

Environmental issues are not always straightforward, however, despite policy requirements. As the previous excerpt shows, by arranging space in different zones, some issues become *problems of a different nature*. An interesting example of how the ecosystems permeate and overflow the management plan's borders can be found in a brief report from the Reference Group's first meeting that was part of the process leading up to the programmed revision of the management plan in 2010. While the Management Forum, the Advisory Group on Monitoring, and the Forum on Environmental Risk Management (headed by the Norwegian Polar Institute, the Institute of Marine Research, and the Norwegian Coastal Administration, respectively) were established to carry out the implementation of the management plan, the Reference Group represented different interests and stakeholders (Ministry of the Environment 2011: 20). Accordingly, the Reference Group worked as a link between a large number of organizations and the inter-ministerial steering committee[9] during the management plan's updating process.

During a meeting that took place in June 2007, some of the participants discussed the geographical delimitation and the management plan's field of action. The reason was that the management area begins one nautical mile from the coastal baseline; however, some seabirds that are included in the plan breed on one side of the baseline, but they feed and winter on the other side; the same thing happens with several fish stocks.[10] In other words, species literally fly or swim in and out of the demarcated plan territory, as their "territory" is more of a territory *in flux*, depending on a wide array of circumstances, such as feeding possibilities, temperature variations, seasonal cycles, external stress factors, etc. According to Tsing, the concept of *assemblage* can help us overcome the idea of ecosystems as fixed and bounded communities (2015: 22). Instead of closed and delimitated geographical territories, which are usually the norm when dealing with political maps or plan areas, we can think of ecosystems as much more open and fluid spaces. While nature's movements, disruptions, variabilities, and shifts are not always easy to take into account from an administrative point of view, they can, nevertheless, be of great importance for human activities and the way these activities are coordinated and carried out. In the case of the Lofoten Islands, the cod's migration from the Barents Sea has been a decisive factor for human settlement, social organization, and value creation in the region. Consequently, many local stakeholders fear that seismic surveys or major oil spills could potentially disrupt or interfere with the way humans and nonhumans have assembled their ways of living over the centuries. The following account by a Lofoten informant emphasizes the importance of the fisheries as the result of seasonal migration and the interplay between species in the waters off Lofoten:

> It remains there [the Barents Sea] until it is 6 years old. Then it is 4–6 kilos. It comes yearly or every other year. It comes to Lofoten when it is ready to spawn. In addition, you also have pollock and haddock coming here, plus other fish stocks that can be on feeding migration, looking for food outside Lofoten. Norway's most important fish stocks, cod, herring, and pollock, they all come to Lofoten and have this as an important area. The cod fishery is the foundation for the settlement in Lofoten. That is why people live here.

With an ecosystem-based approach towards ocean management, the plan process considered the complex interactions and the interrelatedness of entities that produce the marine environment. Based on scientific assessment, some areas within the management plan territory were classified as particularly valuable and vulnerable due to a combination of characteristics and qualities. Most importantly, these areas are "of great importance for biodiversity and for biological production in the entire Barents Sea–Lofoten area" (Ministry of the Environment 2011: 22). The areas that currently have this status and, therefore, require special caution are the Lofoten Island to Tromsøflaket, the Tromsøflaket bank area, including the edge of the continental shelf (Eggakanten), and a 50-km zone outside the baseline from Tromsøflaket to the Russian border. Other parts with the same status are the marginal ice zone around Svalbard and Bjørnøya and the polar front (see Figure 3.1).

As I have indicated already, the environment's vulnerability and value are important arguments used by both sides in the petroleum debate in an attempt to settle Lofoten's future. However, the practical meanings of these categories are negotiated among stakeholders and, thereby, constitute an important strategy used to support political positions and standpoints. The content of these categories is not so much part of the discussion, as most actors and stakeholders understand scientific advice as "indisputable" and, therefore, an important ally in the ongoing political debate. More specifically, what is being negotiated is what kind of *matters of concern* emerge and are enabled by the scientific status as *particularly valuable and vulnerable*. In short, how do these categories politically open up or close the area? From this point of view, "valuable" and "vulnerable" are contested, open, and unsettled categories that have been translated in the ongoing debate into a matter of environmental risk and hazardous gambling, with rich fishing fields, on the one hand, and a matter of strict safety measures, oil preparedness, and technological solutions, on the other hand.

The same areas that are inscribed in the plan document as highly sensitive also participate in other overlapping territories. In fact, the mapped territory is crisscrossed by various "sectorial territories" that incorporate interests and stakes that the management plan aims to coordinate and govern. This situation can be understood as an example of what Law et al. (2013) refer to as *syncretic modes of ordering*. The conflicting interests and sectors operating in the Barents Sea–Lofoten area made the government recognize the need to coordinate both the environment and the interactions between competing industries. The management plan can be seen, therefore, as an attempt to *domesticate* the territorial conflict through scientific advice and spatial organization. As a syncretic mode of ordering, domestication feeds off practices that are *noncoherent*. Accordingly, it is a mode of purification that requires impurities, messiness, and practices that are "unfitting" in order to work and perform (Law et al. 2013: 180). Qualitative differences are acknowledged as such but are then domesticated through a set of multilayered material practices to make them "commensurable and turned into quantitative differences" (Law et al. 2013: 180). However, commensurability is not a naturally given condition or something that is originally there from the outset. On the contrary, commensurability must be produced, and this process is never neutral but is tied to a specific objective or aim (Stengers 2011: 55). Making conditions or circumstances compatible or comparable implies accepting the need for a common standard. From the beginning of the management plan process, the importance of reaching a consensus on controversial issues was highlighted:

> The main aim of the plan is to help achieve consensus among different trade and industry interests, local, regional, and central authorities, environment protection organizations and other stakeholders on the management of this maritime area in accordance with the principle of sustainable development.
> (Ministry of the Environment 2002: 20)

Whether there should be petroleum activity in areas that are environmentally vulnerable or not was addressed primarily as a matter of knowledge. Several uncertainties were translated into indicators, calculations of risk, and the accumulative effects of total human activity on the marine environment. Similarly, the discrepancies initially recognized between the various actors and stakeholders in the planning process were largely *tamed* due to the overarching consensus-based focus that encouraged participants to reach an agreement regarding the scientific standards, procedures, and type of knowledge that should be prioritized.

Consequently, domestication is a common way of ordering and, thereby, dealing with political difference (Law et al. 2013: 188). In short, in political decision-making and environmental policies, controversies are often settled with the help of numbers: majority vote, acceptable levels of risk, cost-benefits, etc. Regarding these matters, the updated management plan states that the scientific documentation on which the plan rests was the outcome of consensus among 26 institutions involved in the planning process (Ministry of the Environment 2011: 20). Overall, the main advantage of focusing on a common agreement regarding the knowledge base was that it made several actors and organizations collaborate across levels and sectors. Conversely, this approach largely downplayed disagreements on scientific facts and the value-laden conflict between use and conservation of the marine environment (Olsen et al. 2016: 300). In the next section, I will analyze how a combination of material practices such as mapping, zoning, and assembling by paperwork (Asdal and Hobæk 2016) simultaneously enacts and domesticates the Barents Sea–Lofoten area.

Mapping vulnerability and risky trajectories

Maps are important governance tools that not only depict a condensed version of space but also produce a territory through spatial organization. As artifacts that arrange and order space, they also comprise a particular form of what I have referred to as modes of ordering (Law 1994). Maps can be seen as spatial artifacts that work as "ordering technologies". In specific located processes, these artifacts become *political technologies*, a term Asdal uses when referring to the various ways scientific knowledge makes its way into the political field and, more precisely, how politics is being shaped and enabled by a series of technical arrangements and procedures (Asdal 2011: 211). By establishing boundaries and relations between certain elements or entities, maps impose a particular order upon the space or the territory they are inventing. I am emphasizing with the term "invention" the fact that the delimitation of a territory is arbitrary, as it could always have been different. Hence, Aligica, a theoretician working outside actor-network theory (ANT), emphasizes "one of the most important features of mapping is that it may take place from different perspectives, inspired by different objectives and employing different techniques" (2014: 79). Consequently, there is nothing self-evident about the way territories are bounded or enacted, as it is always the result of an active decision-making process. In other words, the way maps order space based on aims and purposes means they are "shaped by the problems, intentions, and objectives defining the content" (Aligica 2014: 80). Hence, the mapping of the Barents Sea–Lofoten area in the management plan is the outcome of a combination of political, environmental, and economic concerns. The territory displayed in Figure 3.2 is, therefore, a negotiated territory produced to coordinate and balance conflicting positions between the various actors and sectors involved. Maps are, therefore, "performative, participatory and political" (Crampton 2009: 840) artifacts that work in order to hold things together in a designed and constructed unity. Consequently, the process of mapping is by no means a transparent or neutral inscription of space but a translation, which necessarily means transformation and movement. Maps always operate from a particular perspective, which often implies embeddedness in theory, or institutional discourse. Due to the increased focus on the performative nature of maps, there has been a shift away from studying maps primarily as objects towards focusing on mapping as practice, the knowledge systems involved, and the political field of action (Crampton 2009: 840).

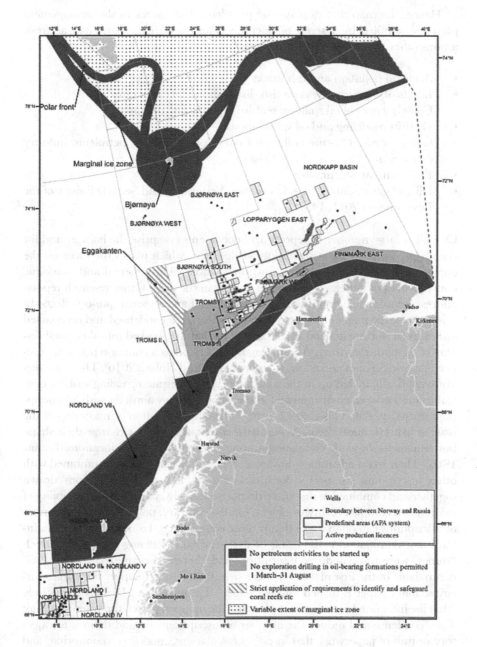

Figure 3.2 Framework for petroleum activities in the Barents Sea–Lofoten area. Source: Norwegian Petroleum Directorate; Ministry of the Environment, 2011.

Hence, the map that displays the petroleum framework in the management plan is assembled by a series of documents and research reports that address, among other things, the following issues:

- Acute oil pollution and fish stocks
- Effects of seismic surveys on fish distribution and catch rates
- Consequences of oil spills on seabirds and marine mammals
- Oil drift modeling and oil spill response
- Consequences of acute pollution from shipping and the petroleum industry on tourism
- Environmental technology
- Oil and gas resources outside Lofoten, Vesterålen, and Senja (Ministry of the Environment 2011: 145)

Overall, a large number of reports and documents comprise the background for the updated management plan's knowledge base, which places the focus on the current state of the environment and the possible environmental and economic consequences of increased petroleum activity in the area. Other research reports address demographic implications, as well as local and regional spin-off effects. In this context, the map of the petroleum framework is a condensed and compacted *assemblage* of documents and research reports that are worked into the visual display of the plan's territory. Assembling by paperwork is a common political practice that brings issues and actors together (Asdal and Hobæk 2016). These matters are synoptically folded up in the map, so to speak, despite revealing striking contradictions between environmental and economic policy aims. As with all inscriptions, maps are flat. This characteristic facilitates domination or mastering of the issue at hand (Latour 1986). As *immutable mobiles*, they do not change their shape but remain stable while circulating among different actors and locations (Latour 1986). Their main advantage, however, is that they can easily be combined with other inscriptions in specific locations or sites. However, inscriptions drawn together and combined in one place do not necessarily guarantee the possibility of mastering a complex reality or object. On the contrary, piling up a large number of inscriptions can often make things more complicated. Latour, therefore, maintains that domination is the result of using the same *flattening* strategy by which "things" were translated into inscriptions and, thereby, became manageable and controllable in the first place. The next step is, therefore, to apply the same strategy so that "paper is turned into *less* paper" (Latour 1986: 21). This is precisely what the integrated management plan and the mapped petroleum framework do: They condense and reduce the number of inscriptions. Behind them is a trajectory or trail of paperwork that inscribes calculations, modeling, monitoring, and assessment of the environment and economic activities. However, this long route of documents is very important because it provides an opportunity "to assemble many allies in one place" (Latour 1986: 23). Consequently, reducing the number of heterogeneous traces and inscriptions from several sectorial agencies and research institutes homogenized the differences, thereby making them tractable. They were *domesticated*, in other words (Law et al. 2013).

Addressing maps as a performative practice is useful because it helps us understand how the mapping of the petroleum framework enacts the territory it is designed to govern. Domestication of the spatial conflict is a double movement in this particular case. First, it reduces and, thereby, makes *flat* the large number of documents and reports by inscribing them into a single graphic, the map. Second, the map of the framework for petroleum activities domesticates the territorial conflict by making use of another syncretic mode of ordering described in the previous chapter, namely *separation*. The map uses temporal and spatial segregation to distribute and organize the territory by establishing a zoning system (see Figure 3.2). Specifically, this means that it employs division to integrate and coordinate the environment and competing industries or activities in the area. In order to produce an integrated plan area, the map divides the territory into several spatial and temporal zones that are designed to balance the controversy by keeping industries, sectorial interest, and stakeholders distributed and, thereby, apart. Accordingly, the map holds things together and performs the territory as "a whole" through *the mode of separation* (Law et al. 2013). The reason for this practice is that the map is embedded in different logics: economic, administrative, environmental, etc. Separation is, therefore, a way to coordinate the territory despite conflicting positions and, thereby, keep *noncoherence* at bay. One important example is the categories, particularly valuable and vulnerable, that are translated into the map as a separate spatial zone, or what we could call a "territory within", where no oil activity is permitted in the present framework. In other areas, both the fisheries and the oil and gas industry are present; however, exploration drilling in oil-bearing formations is prohibited from March 1 to August 31 to protect the fish stocks. Separation through temporal distribution displaces the oil and gas industry during this period from this zone, a decision based on environmental concerns and logic. Syncretic ordering modes can both overlap and include each other (Law et al. 2013: 187). When analyzing the map that depicts the petroleum framework, we see that domestication operates both as a practice on its own and by including the mode of separation as a way to domesticate the political controversy linked to the utilization of the territory. Overall, the mapping of the petroleum framework holds the disputed areas together by a temporal and spatial zoning scheme. More specifically, we could argue that the map works to enact a whole or a unity by not activating all the multiple relationships at the same time. In other words, there must be "territories within" that disconnect spatially or temporarily to achieve a coherent territorial *assemblage*.

However, by translating ecological vulnerability into separate bounded zones, many of the complexities of the marine ecosystems and the multiple relations that make up the status as vulnerable are situated on both sides of the established borders. This shows not only how nature's vulnerability is scientifically being produced but also the way nature is being prepared for political intervention. From an ANT perspective, *"realities are done along with representations"* (Law and

Singleton 2014: 386), which implies that by mapping particularly valuable and vulnerable areas, the categories were also simultaneously being produced. Regarding these matters, a second report from the Reference Group from May 2008 provides important insights as to how vulnerability was constructed during the management plan process. During group work oriented towards obtaining feedback on a draft of a provisional report from the Risk Group (led by the Norwegian Coastal Administration), some participants commented on the importance of clearly defining the concepts of valuable and vulnerable in order to operate with a common understanding. The Risk Group responded that the Norwegian Environment Agency had already started developing these categories, which would provide a common understanding. The report also states that part of this work is to create a visual tool in the form of a map that will simplify the understanding of the concepts.[11] What these brief inputs from the report actually describe is how the scientific categories were negotiated and constructed. Hence, the petroleum framework map plays an important role, as it participates in the process of *flattening* a complex reality and, thereby, making it governable.

Since the map only depicts basic topographic features of the plan's territory while displaying a spatial zoning system, it goes beyond being just a *mimetic* interpretation (November et al. 2010) of the plan area. In fact, the map includes elements and relations that are not based "on some *resemblance* between the map and the territory" (November et al. 2010: 585) but on relations between sectors, industries, humans, and nonhumans, which necessarily imply risk. Although these risky trajectories are not objective features plotted into the map "since risk is not a feature of the 'outside material' world" (November et al. 2010: 593), they are implicitly present in the temporal and spatial zoning system. In other words, the map of the petroleum framework works to anticipate and, thereby, prevent potential risks such as acute pollution and major oil spills. It attempts, therefore, to control not only actors and entities present in the area but, more importantly, the *relations* and *dynamics* between them. Due to certain qualities of the marine environment and the potential risks from maritime transport and the petroleum industry, the map is not primarily concerned with detailed topography but with relations placed in space and time. Risks, therefore, call for a *navigational* approach towards the map instead of a *mimetic* one (November et al. 2010: 582). How to *navigate* risks and conflicts over space in these waters has become a matter of organizing and, thereby, domesticating the potentially *risky relations* between all the actors involved, which means focusing on both human and nonhuman entities. In addition, the map as a material-semiotic reality does not work on these issues in isolation but is part of what Latour (1986) refers to as a *cascade* of other inscriptions. Consequently, it obtains its significance or role as a political tool inside the stream of scientific knowledge that comprises the management plan process. The map's authority is, therefore, referential and interrelated, as it is based on a series of other inscriptions from various locations, including research institutes, directorates, environmental organizations, etc. In the next section, I will further analyze the importance of different knowledge systems as *allies* in the ongoing petroleum debate.

Problematization and enrolment, but where is the obligatory passage point?

As already mentioned, the 2002 white paper, *Protecting the Riches of the Seas*, announced the decision to develop a comprehensive management plan for the Barents Sea–Lofoten area based on an ecosystem approach. The document stated the need to replace the previous sectorial-based management in favor of a new cross-sectorial management regime that could better address the challenges of anticipated increased activity in the area, especially from the petroleum industry and shipping. Consequently, the main concerns in the Barents Sea–Lofoten area management plan are how to achieve coexistence between competing industries and activities while balancing sustainable use and protection of the marine environment.

This is what Callon (1986) referred to as the stage of *problematization*, one of the four moments in his sociology of translation. In the initial white paper, the Ministry of the Environment (and, thereby, the government) defined the issue as a matter of increased economic and industrial activity in the Barents Sea–Lofoten area, something that would also increase the risk of accidents and major incidents. The Ministry of the Environment, therefore, emphasized the need for coordination and spatial planning to prevent conflicts between different users and sectorial interests. This situation also demanded more knowledge about the state of the environment and the total cumulative impact of all human activities occurring in this sea area. Previously, different sectorial activities had been monitored and assessed separately, a situation that had produced a rather fragmented approach towards ocean management in Norway. The main purpose was, therefore, to solve the ongoing tension between use and conservation by adopting an ecosystem-based management regime that could overcome the shortcomings of the previous sectorial-based management.

Problematization is the phase when specific actors try to establish and define the issue and their own role in the network that is emerging around this problem definition. This stage is, therefore, largely about constructing actors and bringing them together. By defining their roles and the relationship between them, actors acquire the capacity to act on behalf of the entire network (Johnsen 2004: 51). From the beginning of the management plan process, the challenges, issues, and solutions were defined, including how to reach the declared policy goals. Accordingly, the plan process emphasized the importance of knowledge-based policy decisions and collaboration between sectors, authorities, and stakeholders to reach consensus. This was important in order to coordinate the type of knowledge that had to be gathered, the actors responsible for producing this knowledge, and the policy areas that this knowledge had to inform. Hence, *interessement* is the process of drawing other actors into the network, so they can fulfill specific roles or positions that have been designed for them. Actors can accept the way they are "locked" into these roles during *problematization* or, on the contrary, they can reject the goals, motivations, interests, procedures, etc., by defining them differently (Callon 1986: 207). In the course of the plan development process, a large number of actors and entities were linked to the policy goals of the new ecosystem-based management regime by making the management plan "also their plan":

> The management plan is also intended to be instrumental in ensuring that business interests, local, regional and central authorities, environmental organizations and other interest groups all have a common understanding of the goals for the management of the Barents Sea–Lofoten area.
>
> (Ministry of the Environment 2006: 15)

The government, through the Ministry of the Environment, also designed its own role by creating the inter-ministerial steering committee and, as a "new" actor, it established an expert group integrating several research institutes, directorates, and government agencies. By organizing conferences and consultation meetings in northern Norway, local and regional stakeholders had the opportunity to voice their position and become part of the network. Conversely, the Ministry of Petroleum and Energy also established a special group to assess how coexistence could be achieved between the petroleum and the fisheries industries (Ministry of the Environment 2006: 16). The actors in this group largely overlapped with the ones in the expert group, which made the network denser with several recurrent interdefinitions and entangled roles. Callon describes *interessement* as a way to gain other actors' interest by producing "devices which can be placed between them and all other entities who want to define their identities otherwise" (Callon 1986: 208). In practice, this means to impede or cut off relations to entities that try to organize or define the issue and corresponding roles differently. In the third stage, *enrollment*, actors accept the roles that have been defined for them, which implies that the network stabilizes as relations become more or less formalized.

The permanent work with the programmed updates of the management plan has managed to largely settle and stabilize the network, whereby expert groups, research and monitoring programs, and various stakeholders have accepted and performed their roles according to how the issue is officially defined and enacted. The final stage, *mobilization of allies*, is a continuous process during which various actors, humans and nonhuman, form alliances and associations as they move around and expand the network. Fish stocks, seabird populations, fishermen, oil resources, experts, stakeholders, local authorities, etc., are represented by documents, research reports, indicators, maps, and seismic data on oil reservoirs, just to mention some examples, in an extended network of translations. The latter are *spokesmen* of the former. The role these representatives have is a matter of continuous negotiation, as there is always discussion "about the representativity of the spokesmen" (Callon 1986: 218). In other words, the social and natural reality emerges from the process of negotiating the various representations involved (Callon 1986: 218). In the case of reaching consensus, the network stabilizes as each entity accepts the role into which it has settled.

Since the Ministry of the Environment, as head of the steering committee, coordinated the stream of knowledge and the roles of the various actors it had created, it is tempting to simply regard the ministry as the *obligatory passage point*.

However, this situation is somewhat ambiguous, at least when focusing on the petroleum controversy surrounding LoVeSe. The whole plan process is primarily concerned with agreeing on the knowledge that is being produced and, more specifically, what kind of political possibilities or limitations this knowledge actually provides. While the intention to create an integrated management plan for the Barents Sea–Lofoten area can be regarded as a shared interest among several directorates, industries, organizations, and stakeholders as it pursued the double objective of facilitating value creation while protecting the environmental value of the area, the act of balancing these two aims is ultimately *a matter of knowledge*. Consequently, knowledge has been translated within the network as a decisive political tool with the virtual ability to settle the petroleum controversy in environmentally sensitive areas. From this perspective, knowledge itself has become the principal mechanism that the petroleum industry must lean on to gain access to the areas that are currently closed to oil drilling. Similarly, keeping the areas closed also passes through the existing knowledge base. In several successive parliamentary periods, opening up or keeping vulnerable areas closed to the oil and gas industry has been defined by politicians as a question of expanding the knowledge base.

While the main political parties in Norway, the Labour Party (*Arbeiderpartiet*) and the Conservative Party (*Høyre*), have traditionally pushed for an impact assessment under the Petroleum Act, neither of these political parties has managed to obtain a majority in general elections during the last parliamentary periods. This situation has forced them to form coalition governments by relying on support from minority parties, which, therefore, have been able to play a central role as the political *gatekeepers* of LoVeSe. In 2019, however, the political scenario changed as the Labour Party announced that it no longer supported an impact assessment that could potentially open up the closed areas outside Lofoten to petroleum activities. While environmental non-governmental organizations (NGOs) have emphasized both knowledge and scientific advice as decisive factors behind this decision, there has never been an official governmental initiative in Norway to close the disputed areas permanently to oil exploitation. Hence, a demand for a permanent resolution will probably be part of the political agenda in the next parliamentary election. Against this background, it is possible to think of *knowledge* as the real *obligatory passage point*. However, the question is about which knowledge or knowledge systems have managed to become indispensable political tools in governing the disputed territory of LoVeSe. In short, where do we find the *obligatory passage point*? In the following section, I will analyze the ambiguous role of knowledge in the production of Lofoten as a heterogeneous territory.

Knowledge systems as governing agents

While the integrated management plan defines the Lofoten Islands as a particularly valuable and vulnerable area and has so far closed it for oil exploitation, this decision is only a temporary oil moratorium. Without a definite political decision yet regarding LoVeSe's future, the question about oil drilling in these waters becomes "a hot potato" on the national political agenda in every new parliamentary period. The management plan, for now, represents a *status quo* for both sides in the oil debate: Those who are against oil drilling have not been able to achieve a parliamentary decision that permanently keeps the oil industry out of the sea area. Conversely, neither have those who welcome petroleum activity been successful in trying to open up the closed parts of the plan's territory. However, as a rolling plan, future updates could potentially change the environmental status of the Lofoten Islands: "This framework will be re-evaluated on the basis of the information available each time the management plan is updated and information from the reports that are to be drawn up from 2010 onwards". (Ministry of the Environment 2006: 127).

This excerpt points towards the provisional character of scientific knowledge. New knowledge can always create new realities and transform old ones. Hence, without enough political support in favor of an impact assessment under the Petroleum Act, the coalition government (the Stoltenberg II Government) intended to fill in some of the knowledge gaps by tasking the Petroleum Directorate to carry out geological surveys that included seismic data from the controversial blocks of Nordland VI, VII, and Troms II (LoVeSe). However, the political situation was very much the same in 2011, as the coalition government was still divided on the Lofoten oil issue. Consequently, instead of giving the green light to an impact assessment, the government initiated a process of knowledge gathering that not only addressed oil-related issues but also focused on the possibilities of value creation and employment within existing industries, such as fisheries, aquaculture, and tourism. As previously mentioned, the management plan stated that this information would be included in the next update of the plan and that it "would be useful if it is decided to carry out an impact assessment of petroleum activities in these areas" (Ministry of the Environment 2011: 135). The knowledge-gathering process was, therefore, tasked as two parallel and competing processes, as the knowledge production was divided according to sectors (Sande 2013: 100). The Ministry of Petroleum and Energy was in charge of researching what in other circumstances would have been labeled an impact assessment, whereas the other four ministries researched the potential for increased value creation and employment within other industries. As a political compromise, the process translates the territorial conflict as a matter of arranging the issues in two separate inquiries, each with a distinct, sectorial focus. This shows some of the ambiguities surrounding the management plan process and the production of knowledge. For now, the management plan definitively works as an important *ally* for those who are against oil activity outside Lofoten, but this situation could change. A politician from the Conservative Party made the following comment regarding the knowledge-gathering process carried out by the previous government:

> They understood that they did not have a majority for an impact assessment. So, they did what they could to prepare in case there was a change after the elections. That is how I see it, and the "knowledge gathering" was sort of interesting. It maps some of the industries and business opportunities in the

region and is, therefore, interesting as such, but there wasn't much new stuff, at least not for someone who comes from the business sector. I didn't feel that many new things came out of it. In my view, they could as well have waited until an impact assessment eventually was carried out.

Another informant regarded the process as an achievement because it provided knowledge in various fields that several actors had requested in order to make an informed decision about Lofoten's future. The informant also emphasized the importance of not labeling the process an impact assessment within the petroleum framework but as knowledge gathering, which put certain restrictions on how this knowledge could be used and what it could do in practice. Just like the majority of those who are against petroleum activity in the north, the informant considers an impact assessment a first step towards accessing unopened parts of the Barents Sea–Lofoten area:

> [We do] not want an impact assessment under the Petroleum Act, because that is the first step towards opening up, and then the government landed on a compromise. They would carry out two "knowledge gatherings" – not an impact assessment. The Ministry of Petroleum and Energy would be in charge of one of them. The "knowledge gathering" regarding petroleum was presented in 2011. It was not an impact assessment, but for those who are positive towards oil activity, this was almost the same as an impact assessment. So, if they decided to carry out an impact assessment, it was almost as it would be ready in two months or something like that. But our victory in this compromise was that there was also going to be "knowledge gathering" about value creation in the north of Norway related to existing industries we have here, industries other than petroleum.

Although this knowledge in its current form cannot give the oil industry access to the unopened parts outside Lofoten, the government could use it as part of an impact assessment if the parliamentary composition changes in future elections. This means that the existing knowledge could be radically transformed if it were translated into other documents within the petroleum framework. By moving from the management plan and the knowledge-gathering process into an impact assessment, this knowledge would obtain a different kind of agency with a completely different field of action. As part of an impact assessment under the Petroleum Act, the current knowledge base will settle into another framework and, thereby, acquire new legal competences, which are key to accessing unopened parts of the plan's territory. Consequently, for the petroleum sector, knowledge is the *obligatory passage point*. However, it is not only a question about the type of knowledge, as much of the information found in the management plan can also be included in an impact assessment. Alternatively, it all boils down to how knowledge is translated and assembled in a specific site and, thereby, acquires *the necessary form to perform*.

Consequently, knowledge that resulted from the management plan process and knowledge as the outcome of an impact assessment are inscribed in different frameworks, which gives them distinct attributes and agencies. We can, therefore, think of the knowledge base on which the management plan rests or knowledge systems in general as what Law and Singleton (2005) call *network objects*. These objects obtain their stable form as an effect of the networks that produce them. Accordingly, knowledge systems depend on relational work to keep them up and going. With reference to network objects, Law and Singleton maintain that "it is necessary to carry on enacting the network of relations that holds them up and constitutes them. Otherwise, things start to lose their shape, lose their characteristics and seep away" (2005: 337). This is key to understanding the ambiguity of the knowledge-gathering process carried out by the ministries: If the knowledge that was produced in the network of relations that emerged around the management plan process is included in an impact assessment, it means that it has lost its original characteristics and shape. Being part of a different framework means that it is enacted by a different set of relations. The displacement from one policy location to another changes its form. However, the networks of relations largely overlap, as the management plan is a cross-sectorial approach and, therefore, includes and works upon the petroleum sector's requirements and needs as well. Against this background, it is possible to think of the current knowledge base as a governing agent that blocks access for the petroleum sector; however, this situation could actually change as the plan process moves through new revisions and updates. For the petroleum sector, not just any kind of knowledge will do the work; it must be formulated as an impact assessment. The knowledge-gathering process can, therefore, be understood as an attempt to locate the *obligatory passage point* outside the constraints of the petroleum framework. However, this outcome could turn out to be only temporary due to shifts in both the parliamentary constellations and the expansion of the knowledge base.

Whereas the majority of Norwegian environmental organizations positively view the decision to gather knowledge without tying it to an impact assessment, some accounts also emphasize that there are certain flaws as to how the knowledge was produced. Regarding these matters, a representative from the World Wildlife Fund (WWF) highlights the lack of "real-life" experiences related to the effects of oil spills:

> There are flaws also in the "knowledge gathering" [process] in which they used modeling to see the different effects oil can have. For example, they have not learned from Exxon Valdez where a whole herring stock was not [made] extinct, but it collapsed because of the oil spills. What the oil spills have done to the ecosystems there was not taken into account as an experience to learn from. They have used data modeling but not actual examples from real life.

A theme that shows up in various accounts provided by the informants in this study is the emphasis on *real-life experiences* as an important variable missing from scientific calculations when assessing human encounters with nature and ecosystems. While the preceding statement highlights the importance of studying real cases to learn about the effects of large-scale oil incidents on the environment, other accounts are more concerned with how some knowledge systems appear as more powerful or display their power when facing knowledge that operates outside standardized scientific procedures. Hence, informants from environmental organizations and the fishing sector emphasized that despite an increasing amount of knowledge on various aspects related to fish stocks and the complexities of the marine ecosystems, this information is often overlooked or downplayed to pave the way for other activities and industries that compete for space. However, both sides in the petroleum controversy emphasized the importance of science as an ally, in other words, "to have science on our side". Making use of facts is, therefore, an important way to substantiate a perspective or position in the petroleum debate. Consequently, informants actively use research reports and plan documents that govern the Lofoten area or sector-related issues to construct their argumentation. This strategy can be observed in the following statement from the representative from WWF:

> We have a report from the Marine Research Institute that indicates that it is an extremely sensitive area in the case of oil spills. So, we should not take the risk – that is really what the government should consider – if they are willing to risk the existing values for short-term profit in a long-term perspective for the oil and gas industry. So, all the gathered knowledge is on our side, we who do not want oil and gas.

Despite scientific advice from several entities, powerful actors like the oil and gas industry are often perceived as capable of outmatching other actors, not only local stakeholders but also acknowledged research institutes. The role of scientific knowledge and the kind of knowledge systems that govern the various economic activities within the Barents Sea–Lofoten area was, therefore, a recurrent theme in many of the informants' accounts. Some specifically focused on what happens when experienced-based knowledge, frequently described as a more subjective knowledge that has been acquired through work and practice, encounters expert knowledge at sea. There are several examples in the data of how expert knowledge does not take into account or directly ignores lived experience. There is the local politician who believes that Statoil (now Equinor) simplified and, thereby, modified the movements of the Gulf Stream in a visual representation of the ocean currents around Lofoten some years ago. According to the informant, the pamphlet the oil company had produced illustrated the current as entering Vestfjorden and then turning out again. There was no information about all the different straits where currents enter that could carry oil spills in case of an accident. He became upset because he knew these streams and currents from working as a fisherman and believed it was wrong to present things in a simplified way "that had little to do with reality". Another account is about the government claiming that there have been few problems related to coexistence between the fisheries and the oil industry in the North Sea. However, the fishermen's experience is completely different, as they have frequently been displaced from sea areas due to seismic acquisition of data. They claim, therefore, that what they experience cannot be labeled as a harmonious coexistence. Other accounts are concerned with how conflicting perspectives represented by different sectorial actors and research institutes "battle" over the right to define the outcome of territorial disputes. The following comment given by a representative from the Norwegian Fishermen's Association illustrates their view of this situation:

> We have experienced that when they are going to issue a license for seismic shooting to survey and explore, it is the Petroleum Directorate that does this. The Directorate of Fisheries in Bergen ... regulates everything that has to do with fish; they give their advice to the Petroleum Directorate regarding whether or not they should open an area and issue an exploration license. The Marine Research Institute does the same thing. Too often, we experience that what the Fisheries Directorate has said in their hearing or in their comments regarding the license is not taken into account and the Petroleum Directorate says yes to go through with it.

This statement points towards one of the main disagreements between the fisheries and the petroleum industry, namely seismic surveys. The reason is that the employment of seismic shooting has been proved to have negative effects on fishing activities and the catch rates as the fish move away from areas where seismic surveys take place. While the startling effect is documented both by research and the fishermen's own experience at sea, there is uncertainty attached to possible long-term effects. Seismic acquisition of data is not just a necessary step during exploration and mapping of petroleum deposits; it is also a requirement for continuous reservoir imaging and monitoring. The fishing industry, therefore, fears that the repeated use of sound waves not only scares the fish but also can negatively influence its spawning and migration patterns. While there have been positive results from dialog between the representatives of the fishing industry and the petroleum sector regarding seismic activity, there have also been situations when the Norwegian Fishermen's Association has had to interfere because permissions for seismic shooting have been granted that they considered would negatively influence their activity.

Heterogeneity and fluctuating risk

A thread running through the empirical data is about the risks related to the possibility of petroleum exploitation outside Lofoten. However, the concept of risk appeared rather elusive, despite the fact that informants talked a lot about its implications. In the beginning, however, I had some trouble trying to understand how risk worked or what it performed in the informants' accounts. There was the idea of vulnerability, of course, but there was also something more to it. Nevertheless, it took me some time to understand that there were several elements and circumstances that went beyond simply linking risk to the idea of a vulnerable nature. After some reflection on the informants' "risky accounts", I realized that I had overlooked the social character of risk. When informants described risk-related circumstances, what they actually described was a series of associations, some of which were perceived as riskier than others. Consequently, we can think of risk as an effect of "a world that is *materially diverse* and *heterogeneous*" (Law and Singleton 2014: 382). This means that risk in this study is produced due to the combination and interaction of a variety of heterogeneous entities, such as a narrow continental shelf, marine species, offshore technology, migration patterns, humans, weather conditions, the strength of materials, etc.

While many informants emphasize that oil extraction is fairly safe by pointing to the fact that there have been few accidents in the North Sea since Norway started petroleum production in the 1970s, other informants reject this perspective. For instance, the following reflections made by a representative from Friends of the Earth Norway:

> It is correct that the possibilities of oil spills are very small, but the consequences would be considerable. That is what risk is all about. You have to risk having oil spills but also the consequences of possible oil spills. That things have gone quite well in the North Sea, well, that is not necessarily what the fishermen say. We know that some of the stocks are declining in the North Sea. The seabird populations are declining along the entire Norwegian coast. It is due to overfishing, but evidently, those toxic discharges we have added to the North Sea through oil and gas production have also had a negative impact on the environment. So, we do not agree with the idea that things have gone so well in the North Sea. On the other side, the lack of space is not as precarious in the North Sea as outside LoVeSe due to the fact that the continental shelf goes further out.

The comment shows that risk is produced by specific associations that differ in the sea area outside Lofoten and the North Sea. The topography of the seabed has very different material conditions that, accordingly, produce different levels and types of risk. On other occasions, risk is produced by entities or factors that have many different locations. These associations link both local and global environmental risk. This is the case of climate change that poses added challenges and elements of risk to those that already exist in the waters outside Lofoten, as a representative from the fishing industry pointed out:

> The heating of the oceans, also the Norwegian Sea and the Barents Sea, will have a considerable impact on the Norwegian fishing industry. It will influence where the fisheries are located and the possibilities of each stock to reproduce itself, grow, and that we have stocks that are possible to fish. Evidently, just 1–2 degrees heating of the Norwegian Sea and the Barents Sea will have uncertain consequences for many of the fish stocks.

Several accounts highlight that risk was primarily a local matter in the past, but as climate change increasingly became part of the national debate and policies, there has been a movement from being perceived only as a territorial conflict related to petroleum activity and the fisheries to also include climate as a significant variable. This has expanded the issue, as it is located not only in Lofoten but also in many other places outside the area. This movement is emphasized in a comment made by a local politician:

> I think that the resistance against oil activity in Lofoten is also partially related to something that has received increased attention in recent years, namely, the climate issue, which has been drawn into the debate. Many people do not manage to distinguish between the climate and the environmental issue and have turned the opening of Lofoten or not into a question about the green shift and the investment in other energy sources and global problems related to climate, climate objectives, and all those things.

The social character of risk, that is, as an association of a series of heterogeneous elements and entities, also has other implications: Risk is always *fluid* and, therefore, operates without fixed definitions or levels. Certain elements and entities make the possibility of future oil extraction in the closed areas of LoVeSe more or less risky. In other words, there are circumstances that cause some informants to welcome oil drilling, despite recognizing the situation of environmental vulnerability. Risk and vulnerability are, therefore, factors that are continuously being negotiated. One important way to reduce and domesticate risk is through technological solutions. Many accounts emphasize the role of technology as a way to master the various challenges related to associations that produce risk. In this context, risk is translated as an opportunity to continue mastering the oil industry, just like Norway has done since the oil adventure started back in the 1970s. Risk can, therefore, be a positive driver to keep up the tradition of technological innovation.

Whereas some informants in principle favor an expansion of the oil industry in closed areas and emphasize the importance of an impact assessment, they also consider that if it turns out that the environmental consequences are considerable due to high levels of risk, LoVeSe should remain closed indefinitely. On other occasions, risk is a *fluctuating* dimension in the encounter with the possibilities of local employment and spin-off effects. However, the most important factor that reduces the perception of risk is the possibility of having infrastructure and technology that allow the oil companies to bring the petroleum resources to land, as this situation would create increased economic activity at the local and regional levels. If this is not the case, the levels of risk increase, even among strong advocates such as this local politician who wants to open Lofoten to oil drilling if national authorities can guarantee that the local community will benefit from the presence of the oil industry:

> My objection is that there has been very little done to study – well it is a political question – what we can gain from it … What can the local community gain from this? But that is more a political question when we get the impact assessment and they study these things. The government and the parliament have to come with some demands regarding local spin-off effects. It has been

my condition the whole time in order to be on the yes side. If we are going to take the risk, we need to be certain about having local spin-off effects.

Risk is, therefore, related to both socioeconomic factors and environmental conditions. Consequently, the management plan territory with its risky trajectories comes into being by a series of overlapping network translations: What on the map is translated as a particularly valuable and vulnerable zone (LoVeSe) is also translated into the petroleum map as Nordland VI, VII, and Troms II. The interplay between these two translations creates tensions and contradictions. We can, therefore, think of risk not only as probability caused by intricate associations in space and time but also as a relational dimension of politically and environmentally overlapping territories.

Notes

1   The marginal ice zone is a transitional zone between open sea and dense drift ice that spans 15% to 80% ice concentration. The southern border of this zone is referred to as the ice edge (Norwegian Polar Institute 2020).
2   Arctic Frontiers is an international conference with a focus on sustainable development in the Arctic.
3   Integrated management plans for the Norwegian Sea followed in 2009 and for the North Sea and Skagerrak in 2013.
4   While the 2011 update of the Barents Sea–Lofoten management plan emphasized the LoVeSe area, the 2015 update focused on the northern part of the plan area/Barents Sea and the marginal ice zone's delimitation. The latest revision from 2020 locates the ice edge slightly farther south than what was the case in 2015.
5   Minister of Petroleum and Energy from 2013–2016.
6   The People's Action for an Oil-Free Lofoten, Vesterålen, and Senja is a politically independent movement, founded in 2009. The movement's main objective is to maintain the blocks of Nordland VI, VII, and Troms II closed to oil and gas activity. Folkeaksjonen Oljefritt Lofoten, Vesterålen og Senja (2015): http://folkeaksjonen.no/content/om-folkeaksjonen.
7   The process ended in a common report: *Future North: Final Report from "Knowledge Gathering – Value Creation in the North"* (2014).
8   Mature fields are defined as fields where the geological conditions are known and infrastructure is well developed. Additionally, these fields have declining production, and more energy is required to extract the oil resources, which increases production costs and carbon dioxide (CO_2) emissions (Arbo 2010: 38).
9   Rapport: 1. møte i Referansegruppen (2007): www.npolar.no.
10 Ibid.
11 Rapport: 2. møte i Referansegruppen (2008): www.npolar.no.

References

Aligica, P. D. (2014). *Institutional Diversity and Political Economy: The Ostroms and Beyond.* New York: Oxford University Press.
Arbo, P. (2010). En næring til begjær, en næring til besvær. In: P. Arbo, & B. Hersoug (eds), *Oljevirksomhetens Inntog i Nord: Næringsutvikling, politikk og samfunn* (pp. 15–44). Oslo: Gyldendal Akademisk.

Asdal, K. (2011). *Politikkens Natur-Naturens Politikk*. Oslo: Universitetsforlaget.

Asdal, K., & Hobæk, B. (2016). Assembling the Whale: Parliaments in the politics of nature. *Science as Culture*, 25(1), 96–116.

Asdal, K., & Moser, I. (2012). Experiments in context and contexting. *Science, Technology, & Human Values*, 37(4), 291–306.

Callon, M. (1986). Some elements of a sociology of translation: Domestication of the scallops and the fishermen of St. Brieuc Bay. In: J. Law (ed), *Power, Action, and Belief: A New Sociology of Knowledge?* (pp. 196–233). London: Routledge & Kegan Paul.

Crampton, J. W. (2009). Cartography: Performative, participatory, political. *Progress in Human Geography*, 33(6), 840–848.

Emmerson, C. (2013). The Arctic: Promise or peril? In: J. H. Kalicki, & D. L. Goldwyn (eds), *Energy Security: Strategies for a World in Transition* (pp. 205–220). Baltimore: Johns Hopkins University Press.

Folkeaksjonen Oljefritt Lofoten, Vesterålen og Senja (2015, July 6). *Om Folkeaksjonen*. Retrieved from http://folkeaksjonen.no/content/om-folkeaksjonen.

Hoel, A. H., & Olsen, E. (2012). Integrated ocean management as a strategy to meet rapid climate change: The Norwegian case. *Royal Swedish Academy of Sciences*, 41(1), 85–95.

Johnsen, J. P. (2004). Latour, natur og havforskere: Hvordan produsere natur. *Sosiologi i Dag*, 34(2), 47–67.

Knol, M. (2010a). Constructing knowledge gaps in Barents Sea management: How uncertainties become objects of risk. *Maritime Anthropological Studies*, 9(1), 61–79.

Knol, M. (2010b). Mot en miljøorientering i planlegging av petroleumsvirksomheten? Integrert forvaltning av Barentshavet og havområdene utenfor Lofoten. In: P. Arbo, & B. Hersoug (eds), *Oljevirksomhetens Inntog i Nord: Næringsutvikling, politikk og samfunn* (pp. 240–255). Oslo: Gyldendal Akademisk.

Knol, M. (2010c). Scientific advice in integrated ocean management: The process towards the Barents Sea plan. *Marine Policy*, 34(2), 252–260.

Latour, B. (1986). Visualisation and cognition: Drawing things together. In: H. Kuklick (ed), *Knowledge and Society Studies in the Sociology of Culture: Past and Present* (pp. 1–40). Greenwich, CT: Jai Press.

Latour, B. (1993). *We Have Never Been Modern*. Cambridge, MA: Harvard University Press.

Latour, B. (2005). *Reassembling the Social: An Introduction to Actor-Network Theory*. New York: Oxford University Press.

Law, J. (1994). *Organizing Modernity*. Oxford: Blackwell.

Law, J., & Singleton, V. (2005). Object lessons. *Organization*, 12(3), 331–355.

Law, J., & Singleton, V. (2014). ANT, multiplicity and policy. *Critical Policy Studies*, 8(4), 379–396.

Law, J., Afdal, G., Asdal, K., Lin, W.-y., Moser, I., & Singleton, V. (2013). Modes of Syncretism: Notes on noncoherence. *Common Knowledge*, 20(1), 172–192.

Ministry of the Environment (2002). *Report No. 12 to the Storting (2001–2002), Protecting the Riches of the Seas*. Retrieved from https://www.regjeringen.no/en/dokumenter/report-no.-12-to-the-storting-2001-2002/id195387/.

Ministry of the Environment (2006). *Report No. 8 to the Storting. Integrated Management of the Marine Environment of the Barents Sea and the Sea Areas off the Lofoten Islands*. Retrieved from https://www.regjeringen.no/globalassets/upload/md/vedlegg/stm200520060008en_pdf.pdf.

Ministry of the Environment (2011). *Meld. St 10 (2010–2011). Report to the Storting First Update of the Integrated Management Plan for the Marine Environment of the Barents Sea-Lofoten Area*.

Retrieved from https://www.regjeringen.no/contentassets/db61759a16874cf28b2f074c9191bed8/en-gb/pdfs/stm201020110010000en_pdfs.pdf.

Nærings-og Fiskeridepartementet, Kommunal-og Moderniseringsdepartementet, Klima-og Miljødepartementet (2014). *Future North. Final Report from "Knowledge Gathering-Value Creation in the North"*. Retrieved from https://www.regjeringen.no/contentassets/8aa6fc353593499ea9e1a343fcb19600/final-report_future-north.pdf.

Norwegian Petroleum Directorate (2010, December 15). *10 Commanding Achievements*. Retrieved from https://s3.amazonaws.com/rgi-documents/e3cbbfde7c90c60753b477e84627ee06dd50ae25.pdf.

Norwegian Polar Institute (2007, June 11). *Rapport 1. møte i Referansegruppen: Helhetlig forvaltning av det marine miljø i Barentshavet og havområdene utenfor Lofoten*. Retrieved from http://www.npolar.no/npcms/export/sites/np/no/arktis/barentshavet/forvaltningsplan/filer/RApp-RefGr-01.pdf.

Norwegian Polar Institute (2008, May 7). *Rapport 2. møte i Referansegruppen: Forvaltningsplanen for Barentshavet og havområdene utenfor Lofoten*. Retrieved from http://www.npolar.no/npcms/export/sites/np/no/arktis/barentshavet/forvaltningsplan/filer/RapportReferansgruppen2008-1.pdf.

Norwegian Polar Institute (Updated 2020, April 23). *The Marginal Ice Zone*. Retrieved from https://www.npolar.no/en/themes/the-marginal-ice-zone/.

November, V., Camacho-Hübner, E., & Latour, B. (2010). Entering a risky territory: Space in the age of digital navigation. *Environment and Planning D: Society and Space*, 28(4), 581–599.

Olsen, E., Holen, S., Hoel, A. H., Buhl-Mortensen, L., & Røttingen, I. (2016). How integrated ocean governance in the Barents Sea was created by a drive for increased oil production. *Marine Policy*, 71, 293–300.

Rommetvedt, H. (2014). Oljeleting Nord for 62. Breddegrad: Norsk Oljemuseum Årbok 2014. Retrieved from http://www.norskolje.museum.no/wp-content/uploads/2016/02/3512_3eeds33e6e6540689b8058ff68abe004.pdf.

Ryggvik, H. (2009). *Til Siste Dråpe: Om oljens politiske økonomi*. Oslo: Aschehoug.

Ryggvik, H., & Kristoffersen, B. (2015). Heating up and cooling down the petrostate: The Norwegian experience. In: T. Princen, J. P. Manno, & P. L. Martin (eds), *Ending the Fossil Fuel Era* (pp. 249–275). Cambridge, MA: MIT Press.

Sande, A. (2013). *Slaget om Lofoten: Olje eller Verdensarv?* Oslo: Akademika Forlag.

Stavanger Aftenblad. (2015, January 20). *Solberg: Iskanten har flyttet seg selv*. Retrieved from http://www.aftenbladet.no/energi/Solberg---Iskanten-har-flyttet-seg-selv-3613200.html.

Stengers, I. (2011). Comparison as a matter of concern. *Common Knowledge*, 17(1), 48–63.

Tsing, A. L. (2015). *The Mushroom at the End of the World: On the Possibility of Life in Capitalist Ruins*. Princeton, NJ: Princeton University Press.

Young, R. M. (1992). Science, ideology and Donna Haraway. *Science as Culture*, 3(2), 165–207.

4   The contested oil of the Ecuadorian Amazon

This chapter presents Ecuador's oil experience, combining the *dependency* and *resource curse* perspectives. These approaches comprise important aspects of how informants frame the outcome of nearly 50 years of petroleum production and exports. The ordering modes that I examine later in this chapter, in one way or another, work upon or against the country's recent economic history, thereby becoming both the material and the frame for ordering. I then analyze the socioenvironmental conflicts of the Ecuadorian Amazon, focusing on the spatial dynamics that arise from horizontal and vertical territorializations. The final section of the chapter looks into the inscription of nature in the Ecuadorian Constitution and the legal implications of granting nature her own rights.

Ecuador's entry into the international market as an oil-exporting country dates back to the early 1970s, like Norway; however, the oil experiences in the two countries have led to very different outcomes. While petroleum exports have largely contributed to Norway's status as a wealthy nation with high living standards for the population at large, Ecuador is struggling with some of the economic pathologies associated with petroleum or mineral wealth. As observed in white paper 25 (1973–1974), the Norwegian government saw its oil resources as an opportunity to substantially transform society. As part of this political project, emphasis was placed on socioeconomic priorities such as redesigning policies of redistribution and social equality, the creation of employment, especially for groups excluded from the labor market, and the expansion of the welfare state. In *The Paradox of Plenty*, Karl (1997) describes how the public sector expanded along with higher levels of consumption, which translated into considerable inflation. Furthermore, a series of symptoms related to Dutch disease also became evident as the contribution of the non-petroleum sectors to the gross domestic product (GDP) contracted, and Norway started to lose general competitiveness due to the high costs of labor mainly caused by an expanding offshore petroleum industry (Karl 1997: 214).

Compared to what happened in other oil-producing countries that experienced similar booms, the political and economic consequences were less severe in Norway than could have been expected. Karl attributes this situation to the fact that when oil was discovered on the Norwegian Continental Shelf (NCS) in the late 1960s, Norway was already a solid democracy with strong public institutions (1997: 216).

This situation provides an important contrast to what has often been the case for developing countries that are oil producers. In Norway, the international oil companies that wished to operate on the NCS had to negotiate with highly prepared civil servants and a well-functioning bureaucracy that, despite lacking experience related to petroleum production and oil concessions, had plenty of experience in long-term planning and administration of public resources in other fields (Karl 1997: 216–218). Moreover, by applying firm policies aimed at controlling public expenditure and borrowing, the country was able to reverse the initial effects of the oil boom and avoid the vicious cycle of foreign debt that has been so common in other resource-rich countries. As opposed to the so-called rentier states, that is, states that live "from the profits of oil rather than the extraction of a surplus from its own population" (Karl 2007: 7), the Norwegian government "resisted the strong temptation to permit oil revenues to replace its normal revenue base by lowering taxes" (Karl 1997: 220). This was important since many oil exporters have experienced fiscal control erosion as oil revenues tend to take over, a situation that, over time, generates strong oil dependency and vulnerability to price fluctuations on the international market. Overall, the capacity to handle the boom effects that follow from sudden wealth gained from petroleum or mineral resources seems to be directly related to the robustness of the state and its institutions. Norway's strong democratic tradition was, therefore, key to the country's performance when faced with the international petroleum industry's challenges and the unexpected opportunities that came with the new oil wealth.

In Ecuador, petroleum resources were discovered in the Amazon region in circumstances that starkly contrast with what happened in Norway, as Ecuador lacked the democratic stability and the strong institutional framework previously described. Although exploration activities initially started on the Pacific coast (Santa Elena Peninsula), where petroleum reserves were found over a century ago, it was not until the 1960s that several international oil companies intensified the search for petroleum deposits in the Ecuadorian Amazon as part of a campaign to increase and secure access to new supplies around the world. During this process, "Ecuador invited foreign capital to invest in petroleum exploration under flexible business conditions and with access to large territorial concessions. The state focused on deriving rents from foreign extraction, hoping more reserves could be identified" (Valdivia 2008: 461–462). While several companies operated in the region during this period, Texaco and Gulf developed the largest oilfields in the northern part of the Amazon basin (Acosta 2009). The oil obtained from the Ecuadorian subsoil generated an atmosphere of general optimism and huge expectations regarding the new possibilities of wealth and development. A former Minister of Petroleum and Energy shared the following anecdote during an interview to highlight the degree of enthusiasm that characterized the first years of the country's oil industry:

> In the late 1960s and the beginning of the 1970s, oil was received with great enthusiasm to the extent that on June 26, 1972, the President of the Republic, the dictator Rodríguez Lara, organized a *fiesta* [party] to move the first barrel

of oil in a parade … a big *fiesta*, and the oil was put in … 20 wooden barrels, to place them in all the provinces, in the prefectures in each and every one of the provinces. One of the barrels was placed at the Pavilion of the Heroes of the military academy Eloy Alfaro, and it is still there. It was a big celebration. It is important to bear this in mind because there was a huge expectation related to the oil bonanza. People believed that oil would solve our problems. There was even an article that I remember from 68' or 69' that I do not have any more, in a Mexican magazine, *Comercio Exterior*, that said "Ecuador: Kuwait of the Andes". So, there was great expectation due to oil, and even in the Amazon, many people had this illusion with oil, and people believed oil would solve all our problems.

The petroleum reserves in the Amazon and the development of an incipient petroleum industry comprised one of the main motives behind the military coup in 1972, which brought to power the Nationalist Revolutionary Government of the Armed Forces led by General Rodríguez Lara (Hurtado 1997: 325). In the previous year, there was an attempt to gain national control over the petroleum resources and foreign participation in activities related to oil operations in Amazonia with the passage of a new law on hydrocarbons under President Velasco Ibarra. As it came into force, the law gave the government an opportunity to review some of the previous concessions that had already been signed with transnational oil companies. Though the new law only applied to contracts signed after October 1971, many of the existing agreements were clearly contrary to national interests. Still, the military government was able to secure a bigger share of the oil rent for the Ecuadorian state with the new legal dispositions (Acosta 2009: 39–40).

During the same period, the state-owned oil company, Corporación Estatal Petrolera Ecuatoriana (CEPE) was established. In 1974, the company acquired a block of shares from Gulf, which led to the formation of the CEPE-Texaco-Gulf consortium, in which the foreign companies held 75% and the national oil company held 25% of the shares. Two years later, CEPE bought Gulf's remaining participation, thereby gaining the majority within the consortium with 62.5% of the shares.[1] These events clearly show that there was a strong element of resource nationalism in the petroleum policies in the 1970s as the military government pursued control over the country's oil reserves and over the international oil companies' participation in the country's extractive economy. Ecuador's membership in the Organization of the Petroleum Exporting Countries (OPEC) in 1973 is significant against this background because it was responding to a strategy of seeking resource sovereignty.

Later on, however, this nationalist position changed as the country entered a cycle of growing foreign debt that was addressed through a series of structural adjustment programs that the government applied under the guidance of international organizations. As part of these economic policies, international financial institutions "pushed for the deregulation and privatization of the petroleum industry to allow the re-negotiation of Ecuador's international debt" (Corbo 1992 in Valdivia 2008: 463). As Ecuador became increasingly dependent on revenues from oil exports and needed to attract foreign capital and new investments, the government modified the previous concessionary policies in a way that clearly favored the private oil companies at the expense of Petroecuador (previously CEPE) (Acosta 2009). The country largely abandoned the previous nationalist approach from the early years of the petroleum industry by adopting a more liberal oil policy.

Towards an oil-dependent economy

In the 1960s, Ecuador was mainly an agrarian society with a large rural population. Political and economic leaders emphasized the need to find a mechanism that could foment industrialization (Little 1992: 46). The oil from Amazonia was seen, therefore, as the answer to this objective, as it both attracted foreign investments and provided the country with important foreign currency. The oil industry generated rapid and partly unexpected economic growth: In 1973, just a year after Ecuador began exporting oil, the prices on the international market multiplied due to the oil embargo and the global energy crisis that followed (Little 1992: 47).

During the oil boom years, the total exports went from approximately 190 million dollars in 1970 to 2.5 billion dollars in 1981 (Acosta 2009: 40). Similarly, petroleum went from representing 14% of total exports in the period 1972–1975 to 70% in 1980–1984 (Quevedo 1986: 94). Needless to say, the injection of fresh financial resources had a huge impact on the national economy and Ecuadorian society. While several sectors clearly benefited from the export-led oil economy, it particularly meant improved living conditions for a growing urban middle class; other sectors did not participate in this new prosperity. On the contrary, the rural population and the indigenous communities were more or less excluded from the benefits of the oil rent. Contrary to what many people experienced in the cities, people's living conditions in rural areas were negatively affected due to stagnation in the agricultural sector, a situation largely produced by economic mismanagement and a lack of investment (Quevedo 1986: 94).

Despite significant flows of economic resources, at a certain point, the oil revenues were insufficient to finance the growing consumption expenditure, especially in the public sector (Quevedo 1986). The government responded by borrowing even more money. Due to the liberalization policies of financial institutions at the time, it was rather easy for the Ecuadorian government to obtain new credit. The reason for this situation was that international financial institutions were holding large amounts of financial resources that they were unable to invest in developed economies due to the ongoing recession. This situation made them turn towards developing economies in the periphery of the global economic system (Acosta 1995, 2009). The petroleum reserves in Amazonia also had the ability to attract foreign capital and lines of credit in a way that the country had never experienced before. In 1984, the foreign debt had reached 7 billion dollars, meaning that it grew 22 times in the period 1972 to 1984 (Quevedo 1986: 95).

Much optimism characterized the first years of the oil boom as people believed the oil rent would finance infrastructure, technology, diversification of the industrial sector, and poverty alleviation programs; in short, the big development push that the country had waited for. This did not occur. Just like the cacao boom, the banana boom, and other booms, the oil boom did not cure the economic pathologies related to underdevelopment that the country had experienced. More oil money did not improve the state's accounts since more income led to increased public expenditure, which generated a constant situation of imbalance and fiscal deficit.

Even with a GDP growth rate of 9.3% between 1972 and 1980 (Quevedo 1986: 94), many sectors did not experience significant improvement nor benefit from the petroleum economy. An important reason was that the structure of Ecuadorian society was not significantly challenged; on the contrary, the property structures remained largely the same, concentrated in a few hands. Overall, the developmentalist approach towards growth did not address class inequalities nor concentration of political power, which continued undisturbed during the neoliberal period that followed the economic crisis of the 1980s (North and Grinspun 2016: 1491).

With falling oil prices, it became clear that the oil boom was over, and those who had been excluded from economic growth and welfare lived in the same or even worse conditions than before. After over a decade of petroleum production in the Amazon, the environment also started to show signs of degradation: Intensive rates of deforestation, contamination of rivers and soil, not to mention the health issues and socioenvironmental conflicts over territories experienced by the local communities due to the presence of the oil companies, especially Texaco.

The resource curse

Many of the circumstances previously described exhibit several characteristics related to *the resource curse* or *the paradox of plenty* (Sachs and Warner 1995, 2001; Karl 1997; Ross 1999; Stevens 2003; Davis and Tilton 2005; Acosta 2009). This literature analyzes why a negative relationship apparently seems to exist between resource abundance, especially mineral or petroleum resources, and the growth rate in many developing countries. These countries seem "to have experienced a worse performance in terms of economic progress and poverty reduction than countries without such apparent 'benefits'" (Stevens 2003: 4). Consequently, this approach represents an alternative view to the conventional position that suggests, "mining plays an important role in the development process by converting mineral resources into an output that can be directly consumed or converted into another form of capital that raises future output in other sectors" (Davis and Tilton 2005: 235). While the "curse thesis" tends to emphasize that developing, resource-rich economies tend to underperform with respect to economic growth when compared to developing economies that lack mineral or petroleum resources, this situation is not viewed as deterministic. However, several authors, including Sachs and Warner (2001), argue that even if resource-poor countries do not always outperform resource-abundant countries, the authors nevertheless maintain that empirical studies show that "high resource intensity tends to correlate with slow growth" (2001: 828). Nevertheless, authors like Davis (1995) reject these findings, arguing that many mineral exporters perform just as well as states that do not rely on mineral exports (Ross 1999: 300). Consequently, the overall purpose of the resource curse research has been to review empirical findings to discover the reasons or driving forces that affect the performance of resource-rich countries and provide a plausible explanation for the curse hypothesis.

In the case of Ecuador, the crisis in the 1980s was, as we have seen, closely related to a series of economic pathologies that originated in the 1970s during the oil boom. The ever-increasing debt was both the cause and the consequence of the continuous flow of financial capital. The debt went from representing 16% of GDP to 42% in the decade from 1971 to 1981 (Acosta 2009: 41), and it was going to get worse: From 1987 to 1991, foreign debt surpassed 100% of GDP (Fontaine 2002: 104). As the state actually spent more than it earned, the rent from the exportable surplus was by no means sufficient to pay international creditors. Ecuador responded by intensifying its oil production without taking into account the amount of proven reserves (Fontaine 2002: 103). The Febres Cordero Administration (1984–1988) simply readjusted the quantity of petroleum reserves to justify the rise in production (Acosta 1990: 236). As it became increasingly difficult to handle the debt service, the country was forced to negotiate the conditions of payment under the guidance of the International Monetary Fund (IMF) and the World Bank. The following structural adjustments had huge social costs for a large part of the population.

Another negative effect of the oil boom was that the economic policies did not manage to produce sustained growth in other non-petroleum sectors; whereas the oil industry attracted labor and investments, agriculture and manufacturing experienced the opposite. This was largely caused by the appreciation of the exchange rate that discouraged non-petroleum exports on the one hand and stimulated an extensive import of consumer goods on the other hand. In other words, Ecuador was experiencing Dutch disease symptoms. It is very common in this kind of situation that an appreciation in the national currency leads to a contraction of the non-resource sector, which tends to be partially replaced by imported goods. This hampers economic diversification and increases dependence on the unstable mineral and oil markets (Davis and Tilton 2005: 236). The appreciation of the currency in Ecuador's case made imported goods more competitive on the domestic market, while the exported products lost competitiveness on the international market (Fontaine 2002: 107).

Furthermore, the mechanism of keeping low domestic interest rates actually discouraged savings and intensified consumption expenditures. During the same period, inflation increased heavily from 9.7% to 48.5% between 1971 and 1990, with peaks around 23% in 1973, 58.2% in 1988, and 75.6% in 1989 (Fontaine 2002: 107). Ecuador adopted the US dollar as its currency in January 2000 in an attempt to stop the high inflation rates and stabilize the economy. However, the process of "dollarization" was socially and economically painful, as it had a huge impact on the cost of living and the competitiveness of the production system (Acosta 2009: 49). Consequently, during the decades of crisis, over 1 million Ecuadorians decided to leave the country. This number is uncertain, as the large majority of this emigration wave were immigrants who entered the US and Spain illegally. However, some people believe that as many as 2.5 million people actually emigrated (Acosta 2009: 49) in order to make a living somewhere else in response to the recession in many sectors of the economy, especially the agricultural sector.

The *curse*, however, was also the outcome of an unstable political system with weak democratic institutions, severe governance problems, and generalized corruption. In a relatively short period, three presidents were removed from office (Abdalá Bucaram in 1997, Jamil Mahuad in 2000, and Lucio Gutiérrez in 2005) due to social unrest and political pressure resulting from their administrations' failures. In this context, it is important to examine the distribution of the oil wealth. In Ecuador's case, it was an *exclusive*, as opposed to an *inclusive*, oil economy, as the oil rent did not actually benefit the communities in Amazonia where the oil came from. Whereas the population in this region directly suffered from social and environmental problems related to the extractive industry that operated there, they were at the same time largely excluded from the economic progress and welfare that other groups in the country experienced. A representative from the Confederation of Indigenous Nationalities of Ecuador/Confederación de Nacionalidades Indígenas del Ecuador (CONAIE) referred to these asymmetries within Ecuador's oil-fueled economic development in the following terms:

> What we have maintained historically is that since Ecuador is a country within a capitalist system, in which the conditions are already given, resources that leave the country benefit certain monopoly groups, so they grow economically. The discourse that dates back to when the first barrel of oil was shipped was that it would eradicate poverty in the country, supposedly from the areas where the oil came from. Well, that has not happened. Ecuador is still poor. Of course, we have highways, of course, there are some people who have become wealthier and other people are poorer. So, poverty has not ended, and the areas where oil is extracted, those are in even worse conditions, as these are the country's poorest provinces.

The situation that the informant describes is in line with what the curse literature maintains about resource-rich countries where local communities have to deal with the environmental impacts and risks that negatively affect their livelihoods and experience higher levels of poverty and exclusion from the national community:

> The exploitation of oil has a profound regional and local impact, and from the standpoint of the majority of the local population, this impact is alarming. Rather than bring prosperity to a region, as is often the claim, the boom-bust cycle associated with petroleum dependence is magnified. Localities where oil is actually located over time tend to suffer from lower economic growth and

lower per capita incomes than the rest of the country, greater dislocations, higher environmental and health hazards, and higher levels of conflict

(Karl 2007: 24)

As the tension between Amazonian communities and the oil companies grew, many civil society groups and environmental NGOs started to request a different kind of approach towards the petroleum resources located in the subsoil of the rainforest. In the following sections, I will analyze how oil as a multiple agent participates in defining and producing space and territory both horizontally and vertically in Amazonia. I also examine the underlying tensions between different understandings of the value of petroleum resources and development. By using an actor-network theory (ANT) approach, I seek to provide another perspective on Ecuador's oil experience.

Modes of ordering the Amazon crude – global connectivity and local fragmentation

To understand how oil works and performs as a multiple agent that produces and connects local and global territories in rather conflicting ways, we must look into some of the implications of objects and realities that are "more than one but less than many" (Mol and Law 2002: 17). Just like the Norwegian oil constitutes a multiple and ambiguous reality, oil unfolds and extends its multiplicity into the Ecuadorian Amazon. The Norwegian and Ecuadorian hydrocarbons are translated in networks that only partially overlap or include each other, despite the fact that they are both commercialized on the international market as energy commodities. Consequently, oil from the North Sea (or the Barents Sea, for that matter) and oil extracted in the Amazon rainforest can never be the same thing: They are different objects since they are enacted in different sociomaterial practices. Simply put, in all these practices, oil "is being *done* or performed differently" (Law and Singleton 2014: 384). Accordingly, oil constitutes a variety of socioeconomic and political objects. It is, therefore, an emergent object that is continuously being reconfigured in new network translations. An illustrative example from the previous chapters is how petroleum reserves currently are being translated as carbon dioxide (CO_2) emissions in various carbon budgets as an essential part of climate accounting practices. As I have discussed already, this situation adds to oil's uncertainty and complexity. The way petroleum associates with the surrounding environment and interacts with technology will necessarily generate distinct types and levels of risk. These factors also contribute to how oil is translated into environmental politics and the national economy.

The materiality of oil

As I discussed in Chapter 2, oil's multiplicity does not necessarily imply a tendency to disintegrate or fall apart (Mol 2002). Conversely, Mol argues that objects multiply along with realities as they are being manipulated in different practices that take place at different sites (Mol 2002: 5). Consequently, to coordinate oil as a coherent reality, it necessarily has to be translated by using some kind of standardization procedure. The API (American Petroleum Institute) gravity[2] serves this purpose as it makes diverse and heterogeneous types of crude oil commensurable. While lighter types of crude oil have an API gravity above 38 degrees, heavier crudes have a density of 22 degrees API or below. Crudes in a range between 22 and 38 degrees API gravity are labeled as intermediate (US Energy Information Administration (EIA) n.d.). A high sulfur content gives a sour crude, whereas lower levels of sulfur make the crude sweet.[3] The specific material conditions or materiality of the *crudo Oriente* and the *Napo*, the two types of crude oil that are produced in Ecuador, become decisive as to how this oil performs and is commercialized on the international market. Its agency is, therefore, produced as a relational effect. In other words, oil becomes an agent, through distribution and interrelatedness, in a rather complex heterogeneous network. Both agency and durability, as is the case with all materials, are consequences of the specific location that an entity holds in the networks of the social (Law 1994).

Consequently, it is important to highlight that the API gravity or sulfur contents are not necessarily significant characteristics in themselves, as the specific material composition of hydrocarbons becomes relevant only when compared to other crude oils on the market or when they are processed as different petroleum derivatives. The price of the *Oriente*, an intermediate, semi-heavy, sour crude with an API of around 23 degrees, or of the *Napo*, a heavy crude with an API between 18 and 21 degrees,[4] is established by using the West Texas Intermediate (WTI), a sweet, light crude with an API of 39.6 degrees (Oilprice.com 2009), as a pricing benchmark. The outcome of this comparison is that the material makeup of the Amazon crude reduces its economic value and, thereby, its financial possibilities. The reason for the higher prices of lighter, sweet crudes is mainly that they require less costly and energy-intensive refining processes (EIA 2017).

A penalty is applied to the exchange value in the market to compensate for the crude's inferior quality. In other words, the Ecuadorian crude has a discount based on the market value of WTI to make up for the difference in quality (Valdivia 2008: 471). Hence, the API gravity works as a *commodification technology*, which allows crude oil with heterogeneous material compositions to become part of, and commercialized on, the same market. As STS scholars have pointed out, technologies are not neutral tools but political achievements with the ability of "reconfiguring boundaries, making connections and creating interoperability where previously there was none" (Müller 2015: 34). That is, the API is a technology designed to overcome and, thereby, *domesticate* and smooth out diversity. Accordingly, the *qualitative* differences of the variety of substances we call oil are translated into *quantitative* economic variations that follow the API classification system. The heterogeneity of oil is acknowledged by using a discount or penalty and is then *tamed* through market mechanisms. In short, variations in material quality are translated into variations in economic quantity, which in ANT terminology *purifies* (Law et al. 2013) oil, as it becomes part of the same system, the international oil market. We can, therefore, think of the API gravity system as part of a set of homogenizing practices on which the oil market depends in order to operate and make tractable the multiplicity of oil. It constitutes what I have previously referred to as a syncretic ordering mode (Law et al. 2013).

Markets, however, are only possible due to a process of framing or disentanglement, which brings agents and goods "into play since all these entities are independent, unrelated and unattached to one another" (Callon 1999: 188). To become part of a market logic, the crude from Amazonia necessarily has to disconnect from a series of relations to establish others. Without this framing process, markets cannot come into being: Their transactions depend on a clear boundary between the actors and entities that belong to the market and those that operate outside its frontiers. Nevertheless, despite disentanglement, the framing process is never complete, as "there are always relations which defy framing. It is for these relations, which remain outside the frame that economists reserve the term externalities" (Callon 1999: 188). These externalities or "rebellious relations", while apparently out of the frame when the *crudo Oriente* has been economically domesticated and coordinated by the international oil market, continue to participate in the local production and negotiation of old and new petroleum territories in the Amazon. It is important, therefore, to take into account that the material specificities of oil have produced and continue to produce spatial organization and territorial relations above the ground and underground. However, while there are multiple studies that focus on the impacts and implications of petroleum territories above the ground, the underground or "the subterranean is an understudied geopolitical space" (Valdivia 2015: 1425). Hence, the focus on the underground as a specific territory allows us to analyze territorial and environmental disputes in the Ecuadorian Amazon as partially the outcome of how specific territories above the surface relate to the underground. Another point of interest is how the extractive activity linked to subsoil petroleum resources and rights territorialize and coordinate relations between the state, various stakeholders, and nonhuman entities above the surface. I will discuss this point further in a later section.

Moreover, the difference in quality that the API gravity system is designed to govern can be attributed to the various geographical locations where oil is extracted from the subterranean. This means that petroleum from different geological reservoirs presents a considerable variation in viscosity, volatility, and toxicity (Oilprice.com 2009). The term "viscosity" describes oil's resistance to flow, as petroleum with a higher density generates friction and, therefore, produces some extra challenges when it comes to extraction, transportation through pipelines, and the refining process (Oilprice.com 2009). These practicalities are directly related to how petroleum and technology interact with each other. An example is Ecuador's oldest pipeline, Sistema de Oleoducto Transecuatoriano (SOTE), that has the capacity to transport 360,000 bbl/d of crude with an API of 23.7 degrees the 497.7 km from the oilfields in the Amazon to the Balao terminal on the coast. However, it is possible to increase its capacity to 390,000 bbl/d by adding a chemical that reduces friction within the pipeline.[5] By making use of diluent agents or drag-reducing agents (DRAs) and, thereby, changing their material composition, heavier crudes can overcome the original resistance to flow. However, high concentrations of sulfur, which characterizes this type of crude, make it corrosive to pipelines and refining technology.[6]

Conversely, volatility is related to how quickly oil evaporates. With highly volatile oils, special measures are taken "to ensure that temperature regulation and sealing procedures lose as little oil as possible" (Oilprice.com 2009). Hence, the mobility and durability of crude oil should be regarded not as inherent characteristics but as part of oil's materiality, and therefore, constitute what Law (1994) refers to as relational effects. Finally, the level of toxicity has to do with how contaminating the oil is when it comes into contact with the surrounding environment and ecosystems. The toxicity of crude oil is directly related to how contamination from oil affects the living conditions for both human beings and nonhuman entities. In the case of oil spills, each crude type imposes certain challenges when it comes to clean-up procedures and environmental remediation (Oilprice.com 2009). This last element is part of what we usually think of as the externalities of petroleum exploitation. It is also important to consider that heavy crudes release more CO_2 when burned than lighter crudes. Although contamination from the oil industry is not directly part of the framing of the oil market itself, the risky associations between oil, human beings, and the environment nevertheless constitute an important part of oil's contingent relationality. In other words, oil is not only ambiguous but also relational as it has the ability to link local actors, ecosystems, markets, and geopolitics in various and conflicting ways. The entanglement of such entities is what makes petroleum simultaneously local and global. Against this background, it is important to emphasize that externalities such as toxic waste, environmental degradation, and severe health issues inflicted on the local population became some of the constitutive elements of the Yasuní-ITT (Ishpingo, Tambococha, and Tiputini) Initiative and part of the ongoing debate regarding *true* development and the value of petroleum resources, which will be further discussed in the next chapter.

Although we tend to think of petroleum as a "natural resource", resource geographers have assertively argued that there is no such thing since resources are always "products of cultural, economic and political work" (Hudson 2001 in Bridge 2011: 821). In other words, resources are made in a series of interrelated and *glocal* processes. The notion of *glocal* or *glocalization* conveys processes that are characterized by simultaneity and the interconnectivity between local and global arenas (Robertson 1995: 30). Another possibility is to think of glocal in a more Latourian sense, which means to abandon the common view regarding the local and the global as preexisting scales of social life and, instead, think of them as emerging outcomes of human and nonhuman interactions. Consequently, from the perspective of a *sociology of associations*, global and local are the effects of ongoing practices (Blok and Jensen 2011). Latour explains his argument in the following way:

> As soon as the local sites that manufacture global structures are underlined, it is the entire topography of the social world that is being modified. Macro no

longer describes a *wider* or a *larger* site in which the micro would be embedded like some Russian Matryoshka doll, but another equally local, equally micro place, which is *connected* to many others through some medium transporting specific types of traces. No place can be said to be bigger than any other place, but some can be said to benefit from far safer connections with many *more* places than others.

(Latour 2005: 176)

Hence, the difference between the global, the national, and the local is not a matter of size but a question of "more or less dense connections" (Müller 2015: 35). When I discuss natural resources as the outcome of glocal processes, I specifically refer to how resources, in order to become such, depend on a chain of actors and entities situated in many different places at the same time. In the case of Ecuador, the oil deposits and the drilling technology are located in the Amazon, but they are also articulated in many other places through oil policies, pipelines, geologists, research, stock markets, oil tankers, refineries, consumers, etc. Without all these *localities*, crude oil would not become a global resource simply because it would lack all the necessary relations and links that collectively generate a resource as a resource. However, the production of natural resources and the glocal dynamic that sustains it is often overlooked and, in some cases, even "black boxed" (Latour 1987). In the case of the Amazon crude, however, the circumstances that I have described – oil's materiality and the extended networks where it is produced – become important elements of contestation.

In the next sections, I will describe how this plays out in three ordering modes that I identified when analyzing the empirical data from fieldwork. These are *developmentalism, destruction,* and *violation of rights* (the rights of nature and humans). While forming distinct recurrent patterns, these ordering modes clearly interact and interfere with each other as they actively attempt to rearrange elements from the other modes as a way to oppose or counteract their story-telling and sense-making. All of them, in one way or another, relate to Ecuador's oil history. The informants, however, arrange this history in very different and often conflicting ways. One way to understand this situation is that Ecuador's history of increasing oil dependency constitutes a common background or context for ordering the role of petroleum resources. However, this context is not passively "lying around", but is actively worked upon, pulled apart, questioned, challenged, embraced, and contested, which indicates that ordering is largely about constructing specific problems and problem solutions (Law 1994: 83). Interestingly, in this specific case, the solution in one ordering mode is translated as a problem in another one. Consequently, they seem to work together somehow in what may be thought of as intertwined oppositions. As I have previously stated, I do not maintain that the ordering attempts I identified in the data material are the only possible modes in the ongoing controversy surrounding petroleum extraction in vulnerable areas, as there are definitely other possible ways to order the Amazon crude. I find it very likely that these modes can change or transform into alternative ordering strategies in the future. Although the ordering patterns that emerged from the interviews could turn out to have more general applicability, they exist neither independently nor free from their local performance and embodiments (Law 1994). This means that they are closely knit together by lived experience and recurrent practices, which provides them with particular temporal, spatial, and material dimensions.

The developmental oil: Past and present

While arranging present-day actors, territories, materials, and policies, *developmentalism* is an ordering mode that is also highly concerned with Ecuador's past. It needs the past to make sense of the present and, therefore, actively works upon the country's oil history, which becomes part of the framing. When this ordering mode looks back, it tells about the colonization of the Amazon in the 1970s and 1980s, a process that involved two very different types of *colonizers* that nevertheless carried out the task as a correlated but not coordinated effort: The oil industry and the *colonos* (settlers). The agrarian colonization largely followed in the footsteps of the oil companies due to the roads they built cutting through the rainforest as they provided access to previously closed territories. Roads meant the possibility of commercializing products by bringing them to markets, something that without access constituted almost an impossibility.

In all three modes of ordering, there is a lot of story-telling about how roads, by opening up the territory, not only changed the state of the environment but also modified cultural practices and social organization in many indigenous communities as the roads brought both connectivity and local fragmentation. Roads became one of the most important infrastructures in the Amazon region, producing increased mobility and means of communication. *Developmentalism* implied that the transportation system comprised an achievement not only for the petroleum industry but also for the inhabitants of the Amazon. In this ordering mode, roads were also an important part of the national development policies, as they integrated the country and linked the national petroleum industry to the international oil market. While this ordering mode stresses the importance of the road system as a mechanism to colonize the territories of the Amazon, the other two modes clearly enact it as an essential and, therefore, problematic component of the oil issue. The reason is that these ordering modes enact the presence of the oil companies and their infrastructure as the spearhead that opened up the Amazonian territories to a massive and disorderly process of migration.

The presence of the national and international oil companies was portrayed, overall, as an important strategy to develop and modernize the country. The output from the petroleum industry generated important income for the state that could be invested in developmental projects considered a high priority. Hence, the military government saw the process of colonization as a way to incorporate and bring the Amazon, the "undeveloped frontier of the nation-state" (Valdivia 2008: 464), into the circuit of the national economy and the political domain. The law for colonization of the Ecuadorian Amazon was passed in 1978 as part of a strategy to *physically* and *politically* take possession of the region and transform it into an economic territory. The law had the character of a national priority, and the state offered land-poor settlers 50 hectares of land with the condition of clearing at least 50% of it within a period of five years. If they were unable to meet the requirement, they risked losing their property rights (Fontaine 2007: 275). Consequently, the migration of land-poor *colonos* who left the coast and the highlands in search of new opportunities was a key element in the military government's scheme to *connect* and *spatially organize* the territory. This also constituted a way to expand the agricultural frontier while simultaneously reducing the pressure on the land in other parts of the country. The *colonos* were also meant to take on the role of "living frontiers" *(fronteras vivas)* based on their declared Ecuadorian identity (Little 1992: 61). The idea of borders and frontiers as a form of national identity that is materially embodied in practice is interesting. This shows that *developmentalism* was preoccupied from the very beginning with organizing and arranging nature, space, and people, and the relations between them, as a form of controlling and domesticating the various ways of life in the rainforest. Similarly, this ordering mode also reveals the importance of identity, particularly the importance of establishing a national identity in the region as a means of ordering the culturally heterogeneous human groups that live within the Amazon basin.

Furthermore, the *developmental mode* enacts identity as something that was not necessarily located in Amazonia but was distributed in other parts of the country and, therefore, had to be imported together with the *colonos* as part of national security measures. Consequently, homogenizing identity politics, represented by the settlers, the oil companies, and the state through its national security policies, worked as a governance tool to manage the complexities and disorderliness of what was regarded as a *savage, uncivilized and uncertain territory*. Reflecting on the state's role in the country's developmental efforts in the past, an informant made the following comments during the interview:

> I believe it is true that was what the world capitalist system imposed at the time to make you depend on capital, to make you always depend on another country and a third party; financial sources, the IMF, the World Bank. It is true, and we have not learned the lesson. We are still engaged in primary exports, and we continue to increase our financial commitments with oil and now also with mines, as we are opening up another front. That is number one, and two, having exploitation systems that are terrible. What happened with Texaco is a world brutality. In addition, Texaco gained a considerable amount of money, which means that there was nothing left for the state, at least very little. Terrible oil deals with an infrastructure and petroleum technology that was a disaster. It was not only because the petroleum technology was different at the time but also because the standards and national control did not exist back then. Texaco did as it pleased. It was an abandoned territory. The state lost. The state did not exist in the Amazon, and that is a real issue and another lesson.
>
> Me: *So, the state was absent in this region?*
>
> Completely absent. Going back to these places is terrible; it is really hard. An absent state but also a homogeneous state that tried to homogenize

everything. I am not talking about Correa,[7] I am taking about the Ecuadorian State in general throughout history.

While the account emphasizes how petroleum exports increasingly shaped Ecuador's role as a dependent actor within the capitalist world-system, or in other words, represented what Wallerstein (2004) and dependency theory refer to as periphery or satellite economies, it also stresses the absence of the state in the Amazon region. Some authors refer to this process as deterritorialization (Gudynas 2005; Acosta 2009); that is, a situation in which the state's capacities are modified, which complicates the regulation of the territory's use, the application of justice, and the ability to handle environmental impacts (Gudynas 2005: 5). The state's reduced capacity also translates into difficulties when it comes to managing the extractive production in the vast territories within its borders (Gudynas 2005).

Although the state has failed to control and regulate the ways in which the petroleum industry has employed the territory, it has been highly active and present in what the informant regards as a process of homogenizing the area. Consequently, in *developmentalism*, diversity can easily become an issue and must be kept within certain constraints. According to Little (1992), within the developmental ideology, the forest became an obstacle or an impediment that was best done away with. That is, the dense and humid jungle had to be conquered and subjugated, which meant abandoning the indigenous ancestral practices related to land and adopting what was considered "modern" agricultural methods (Little 1992: 62). The forest (and its peoples) was standing in the way of development and progress. It goes without saying that this mode performs a *dualist* vision of the world in more than one way: Modernization and development versus traditional ways of life; progress and infrastructure versus uncertain and complex environments; "civilized" national identity versus "savage" local communities, etc. Law (1994) specifically talks about the drive towards *dualism* as inherent to the modern project. In short, this ordering mode largely embodies the foundations of modernity with clear delimitated categories and pure forms, which is the outcome of the continuous struggle to keep nature and culture as two distinct and separate domains (Latour 1993). Both past and new versions of developmentalism or "neo-developmentalism" (North and Grinspun 2016) have ramifications back to the Enlightenment project with its strong focus on the rights of the free individual and the possibility of rationally transforming society to achieve incremental progress. Regarding the philosophical background of developmentalism, in his work on modernity and dependency theory in Latin America, Grosfoguel says that:

> The modern idea that treated each individual as a free centered subject with rational control over his or her destiny was extended to the nation-state level. Each nation-state was considered to be sovereign and free to rationally control its progressive development. The further elaboration of these ideas in classical political economy produced the grounds for the emergence of a developmentalist ideology. Developmentalism is linked to liberal ideology and to the idea

of progress. For instance, one of the central questions addressed by political economists was how to increase the wealth of nations.

(2000: 348)

Ideas about modern secular states as founded on liberal democracy, reason, and empirical science have produced specific conceptions regarding the accumulation of wealth, the market, and the role of the state. During the Enlightenment, the scientific method and ideas about science's ability to acquire knowledge about an exterior nature that could be represented and governed also shaped strong beliefs in the human ability "to organize collective life in stable and transparent systems of relations" (McNeish and Borchgrevink 2015: 17).

Hence, behind the *developmental mode*, we find Boyle and Hobbes, the "founding fathers" of modernity, lurking in the background. Latour (1993) claims that the outcome of their controversy was an ontology that decoupled nature and society and, thereby, the spheres of humans and nonhumans. This fundamental disengagement has developed into rather fixed understandings of what constitutes progress, the creation of value and the common good. In recent years, however, this approach has been increasingly challenged in Ecuador as some political actors and stakeholders problematize the government's conception of development based on what they regard as a "neo-extractivist" model in which nature's value is primarily instrumental as natural resources. Regarding this approach, a representative from Ecuarunari (Confederation of Peoples of Kichwa Nationality) in 2015 argued:

> Then there was a lack of economic resources. Correa's view, of course, is that we cannot live like beggars sitting on a sack of gold. So, he said that extracting [oil] in Yasuní would transform and eradicate poverty in the communities, among the nationalities. He, therefore, took a step away from ecological politics, a green politics, favoring extractivism, a reprimarization of the economy.

Along the same line, an informant from the Yasunidos collective commented about the government's decision to terminate the Yasuní Initiative and begin oil exploitation in the ITT block. The account emphasizes that the government's political goals and the stated developmental agenda are being legitimized by the population's socioeconomic necessities:

> In a country as diverse as Ecuador, diverse in terms of ecosystems, diverse in terms of culture, extractivism is your worst enemy. We have seen that you do not "sow oil". You do not create development this way, not true development, not development through roads but development in the quality of life for the population. Therefore, extractivism is not the way. That is the reason we now not only say no to oil but also to mining because extractivism is not the only option. There are many other alternatives: Tourism, renewable energy, and the field of bio-knowledge. There are alternatives to work together with

nature and not against nature. I think that is the most important lesson we have learned ... but the [discourse] is that Ecuador is a country where there is still poverty, where you have economic needs, development needs and we cannot be so absurd that we do not extract the oil.

The interview excerpt points towards a central issue in the ongoing controversy regarding the relationship between extractivism and development, namely the tendency to leave socioenvironmental costs out of the equation. The developmental ordering mode emphasizes the importance of expanding and intensifying petroleum extraction (and the extraction of natural resources in general) as a means to reduce poverty and pursue social equality for groups that traditionally have suffered exclusion within Ecuadorian society. While the government has clearly dispersed the oil rents through public investment and social programs, many accounts claim that a more profound redistribution is still to occur, as economic and societal structures remain largely the same. In addition, *development* and particularly *developmentalism* contains its own antithesis, as there is considerable tension within this mode that is performed as an explorative search for an *alternative ordering mode*. In the next section, I will discuss how *destruction* and *violation of rights* play an active part as they clearly interfere and collide with various aspects enacted in developmentalism. In other words, they inhabit developmentalism as the moral and legal foundations behind the urgency of finding a different developmental paradigm, which is not based on the "extractive imperative" (Arsel et al. 2016). Whereas some accounts tentatively mention some productive sectors and technical innovations that could, in principle, replace the contribution of the extractive industries, particularly oil exploitation, to the national economy, others stress the need to pursue a different kind of development but are not clear about the alternatives.

There is a lot of story-telling, however, about the current developmental paradigm translated as technocratic and homogenizing policies in which the state, as the guarantor of the political or common good, has overlooked the diversity of cultural identity and sociomaterial practices. Some informants see this as a homogenizing drive represented in the construction of modern infrastructure, such as the "millennium schools" and the "millennium cities" during President Correa's administration (2007–2017).[8] Consequently, the transformation of the state, which has gone from having a weak presence in the Amazon region in terms of education and health to having a strong presence in all fields, is described with ambivalence, as the current social planning "contradicts the reality of the Amazon". This mode of ordering is, therefore, largely preoccupied with performing the distance between how things currently are and how they *might or should be*. Despite incipient debates that envision a different kind of development situated outside the confines of "neo-extractivism", which is believed to better articulate state policies with *true* developmental needs, the discussion also exhibits certain shortcomings. Arsel et al. (2016) have argued that the main weakness regarding the post-extractivist discussions and literature is that they seem to overlook the material and social needs that are currently being financed through revenues from extractive activities such as petroleum production. This situation reduces their policy relevance (Arsel et al. 2016: 883). Nevertheless, within the developmental mode, a variety of proposals and theoretical perspectives that range from debates regarding the good life or "good way of living" (*buen vivir*) to notions about post-development (Endara 2014) have tentatively been drawn together in a search for an alternative mode of ordering development.

In Latin America, developmentalism in its old version is often referred to as *cepalismo* (CEPAL, in English, the Economic Commission for Latin America and the Caribbean; ECLAC). The ECLAC was established in 1948 to foment development in the region by focusing initially on industrialization through import substitution. This meant, in practice, making use of trade barriers and tariffs to protect the domestic market from foreign competition and, thereby, promoting the economy's diversification. The intention, in other words, was to pursue less dependency on global markets to strengthen domestic production. Largely inspired by Keynesian economic theory, the Latin American *cepalismo* emphasized the role of the state in planning, investments, and employment creation. While partially contested by several other approaches aligned with Marxism and dependency theory, the CEPAL policies were nevertheless influential in the region up to the early 1970s. In the wake of the economic recession of the 1980s and the neoliberal economic model that followed in the 1990s, several left-oriented progressive governments implemented what is regarded as a "new developmentalist" approach. Their policies were principally aimed at incrementing the state's social and economic planning capability and securing well-being and public services for large sectors of society financed with income from oil, minerals, or other types of primary exports (North and Grinspun 2016: 1484).

These governments have been fairly successful in targeting poverty and the exclusion of vulnerable groups, yet other policies oriented towards diversifying the production system and transforming the energy matrix have not yielded the same positive results. According to the critics, the main reason can be found in current economic policies that focus on the expansion of extractive industries, that is, by reproducing former patterns of dependency that the *cepalist* policies tried to overcome (North and Grinspun 2016: 1484). Although expanding the extractive frontier is a tendency that can be observed worldwide (also in Norway), the idea of a positive relationship between extractivism and development has proved to be particularly strong in Latin America, where the role played by extractive industries has been "placed at the heart of modern development" (Arsel et al. 2016: 880).

Hence, according to the informants' accounts, *developmentalism* is constantly being performed as a mode that *recycles the past*, yet old formulas, policies, and discourses are not necessarily able to address the current developmental challenges that Ecuador faces. An informant working with environmental issues and communities in the Amazon stated:

> Maybe the former developmentalism had the idea to construct a market disconnected from globalization. I believe this one is oriented towards reinforcing the country's participation in the global market but it is an

"old-fashioned" developmentalism, that is, natural resources to finance development, and I think that the president deep down always had this vision. In other words, I think that all these green economy issues never managed to become part of his repertoire of possibilities ... I think it is problematic that a state lives off oil. Well, more than live off oil that it has a rentier economy, because I think there is a huge difference with Norway. The Norwegians have oil, they depend a lot on oil but it is not the country's only productive sector. They have the whole fisheries system from the past; there is some important industrial development as well. In Ecuador's case, it is a typical rentier economy. To oppose the development of the industry that sustains the Ecuadorian state's economy is impossible. We are not even talking about opposition. Based on my experience, when you ask people in the Amazon, they do not tell you that they are enemies of oil but they do not want that their rights to a life, health, and territory to be violated.

The account addresses the present developmental efforts carried out by the Ecuadorian government as some kind of mimetic practice or continuity with the past that cannot rise to the challenges of present economic and socioenvironmental needs, especially in the Amazon. The critics describe developmentalism using various theoretical perspectives. In this case, concepts about the rentier state, rent-seeking behavior and, more vaguely, the *curse* participate in framing and tying together the state's failure to generate sustained and *true* development. It is within this ordering of materials, agents, and discourses that oil's ambiguity most clearly surfaces, as it is both an agent capable of development but also an agent that, due to its material and financial properties, generates the opposite. As the developmental mode, in its more critical, ideological version, developmentalism, is on the search for alternative modes of ordering, comparing and contrasting become recurrent patterns. In the account above, Norway is being performed as a benchmark to illustrate that it all depends on how the oil economy is *practiced*, so to speak. The ambiguity of petroleum resources is also present in a comment made by a former member of the Administrative Committee for the Yasuní-ITT Initiative:

> Without oil, we would not be where we are [today]. We would find ourselves in a complicated economic situation but we have also learned that the dependency of extractivist countries is a really serious matter and the effect that the extractive system has, which belongs to the state, since the democracies are not very stable ... The real issue is that the government does not need to be accountable to its voters as it says I will give you so much and you give me the votes because it does not depend on taxation from the citizens but on the oil that belongs to the same government.

In *developmentalism*, there is a lot of critical story-telling through theoretical perspectives (which are also ways of ordering). Again, the framing seems to move between the curse hypothesis and the dependency approach through concepts related to the rentier state, which depends on revenues from petroleum exports and not on a regular taxation base from the population. The outcome is a lack of accountability and fragile democratic institutions. Hence, theoretical concepts play a central role in displaying the potential that resides in oil's multiple agency: Reality could always have been different. Consequently, there is a kind of meta-ordering going on in this mode: *Ordering through ordering* as all theories in one way or another attempt to organize the world and tell coherent stories about it. Law says that ordering modes "tell of themselves, they perform themselves, and they embody themselves in different materials" (1994: 151). In this case, the developmental mode seeks representation in theories, which become its "materials of representation" (Law 1994: 151). In short, this ordering mode put theoretical perspectives to work, as they are clearly employed as *spokespersons* to systemize and make sense of the country's oil experience.

I will examine in the next section how *destruction* and *violation of rights* work within and against this ordering mode by pulling it apart and reassembling its bits and pieces to enact Ecuador's oil history not primarily as economic and developmental policies but as lived experiences and socioenvironmental risk.

Oil as destruction and violation of rights

While *destruction* and *violation of rights* constitute two distinct ordering modes, I have chosen to analyze them together since they are closely related and collaborate in their ordering efforts and their opposition regarding developmentalism. A difference between them exists, however, as the first mode revolves around experiences related to oil's devastating capacities, and the second mode is more oriented towards the legal implications of petroleum activity in the Amazon. Whereas the previous ordering mode, developmentalism, orders the role of petroleum resources through extensive use of theoretical concepts, *destruction* tells about oil from the standpoint of situated lived experience. This ordering mode patterns the effects of petroleum activities on the lives of human and nonhuman actors. Hence, *destruction* gets close up on the material conditions of oil, as it tells about collective and individual experiences of living in proximity to the oil companies and their infrastructure: Contamination of rivers and soil, deforestation, species loss, cancer, cultural fragmentation, animal deaths, and intoxicated bodies. These patterns of devastation are displayed by turning to the experiences and localities where the Amazonian population lives and interacts with the crude's material, economic, and political conditioning that shape their livelihoods and their ways of associating with the environment. Hence, this ordering mode is largely *testimonial*, as it gathers and displays learned and embodied experiences from various actors and stakeholders. Regarding these matters, an informant from a family of *colonos* explained how he learned about the effects of crude oil and the presence of the oil company:

> I have lived there for 28 years and have seen the damage caused by Chevron [previously Texaco] directly in the Amazon. I live in the Shushufindi canton

that is part of the province of Sucumbíos. As I said I came 28 years ago, and I also worked 4 years in the oil company. I have also worked a lot with social issues with the priests from the Catholic Church, for example, with whom I visited the communities, and I was able to directly see people's problems related to their health and basic services, such as water, among other things. I have also seen many problems, many friends who have died along the way with cancer. So, these are personal experiences more than experiences people have told me about because I am also part of the story.

This mode of ordering the Amazon crude is largely about learning lessons and, more specifically, painful lessons from the past. It is clearly concerned with enacting a kind of *apprenticeship* regarding socioenvironmental conditions related to the dividing line before and after oil. Consequently, there is a lot of telling about the movement from not knowing about the consequences of oil extraction to acquiring knowledge about the negative impacts from direct and embodied experience. Regarding this increased level of knowledge, a representative from the environmental organization, Acción Ecológica, stated:

I think there has been a lot of learning, unfortunately, regarding the impacts that petroleum activity generates, [and] understanding of how these impacts occur in all their metabolic process[es]. In other words, not only during extraction. Conversely, it starts with the first contacts, with the negotiations, how the companies work, the operators. It is impressive how people in the communities know about petroleum engineering like nobody. I believe that at the national level there is questioning going on. Oil cannot be a country's source of wealth if it brings poverty to the territories. However, I think that in this last period the most important lesson is the opposite. It revolves around what we do not want to sacrifice and I think all this about Yasuní helped a lot because all these images of people with cancer, of wastewater pits, dead animals, they truly generate rejection against petroleum activity but not a rejection with the capacity to mobilize people. Horror does not mobilize people. What moves people is hope, it is the positive thing, the fight for life, for what we still have and I believe this is the big leap.

Destruction does not exclusively tell how the oil industry generated asymmetrical power relations and fragmented the Amazonian communities. It is also a *counter-story* that actively works upon these relations. *Destruction* assembles a different kind of developmental pattern by questioning and rearranging actors, materials, technology, and territories in which ranks and hierarchies are reconfigured to produce a more symmetrical relationship between nature and humans. This ordering mode is developmentalism revisited, in a certain way, as it works to expose this mode's failed strategies. In the following comment about the geographical distribution of poverty, a representative from CONAIE emphasizes the failure of oil-driven development:

Statistics say so, and in the Amazon, we can compare Morona Santiago, which is one of the provinces in better economic conditions, where you do not have petroleum exploitation and you have anti-oil resistance. In areas where there is oil exploitation, there is more poverty. So, that is what we question and it has been a key argument: That the resources that come from this [region] have not been able to solve the country's poverty issue, but certain sectors in the country get rich, certain sectors located in the cities of Quito and Guayaquil.

This account addresses the state's territorial dynamics in the Amazon, which have practically reduced the region to an "enclave" status within the national economy. Throughout history, the Amazonian territory has occupied a position principally as a provider of natural resources and, in the last 50 years or so, specifically as a provider of petroleum and minerals. Despite contributing heavily to the national economy and GDP, the Amazon's economic and political articulation with the rest of the country has historically been based on socioeconomic exclusion. In a way very similar to past colonial, politicospatial organization, the resources from the Amazon have not provided direct local returns. On the contrary, the oil revenues have benefited other regions and sectors of society while simultaneously fragmenting communities and ecosystems locally through roads and transportation systems, contamination, marketization of the economy, and ambiguous relations with the oil companies. In the absence of the state as a guarantor and provider of important services, such as education and health, the population was often forced to negotiate with the companies that came to replace the state in many of its functions. The oil companies addressed collective demands from communities and indigenous organizations through different social programs that frequently operated outside the law and lacked proper planning (Acosta 2009: 156).

Moreover, these projects were designed primarily to pursue "friendly" relations with the local communities and, thereby, they became important mechanisms for reducing opposition and resistance, as the population often ended up depending heavily on the oil companies to satisfy necessities. With the state as a "missing link" in the Amazon, except for the military presence to safeguard the companies' operations, the petroleum sector largely operated in isolation from the rest of the country without major interferences or scrutiny. If not officially, but at least in practice, the state handed over many of its functions to the oil companies, which became powerful actors over time. This situation illustrates some of the economic and political contradictions that this ordering mode works upon and displays. There is certainly tension and contradiction between the local and national effects of the oil-fueled economy. For instance, this ordering mode tells how roads and pipelines became *matters of concern* due to their material and political uncertainty. On the one hand, they comprise the necessary infrastructure for transporting and exporting the crude as they create both national and global connectivity; on the other hand, these entities create physical and social fragmentation at the local level. In practice, this means that some connections are replaced and disappear due to new network translations, while others are considerably weakened in the encounter with powerful and dense interconnectivities such as the international oil market.

In the case of *destruction*, the vision of agency being performed is not primarily economic (as in developmentalism); instead, it focuses on oil's materiality, bringing conditions such as toxicity and viscosity of heavy crude oils to the forefront. Due to *predator* technological and corporative practices, these characteristics become highly problematic when they come in contact with the surrounding environment. In other words, there are no traces of what Law calls "technical heroism" (Law 1994: 129). On the contrary, *destruction* is inhabited by *technical distrust*, which can be observed in the following comment by a member of the Yasunidos collective about the government's promise of employing cutting-edge technology in the Yasuní-ITT oilfields:

> Faced with questioning because of the levels of devastation, contamination of all kinds, the argument they used and ... graphically presented ... is: Technology today has changed so much that there will be a minimum impact. There will not be significant environmental damage, and they are going to be environmentally responsible, which is supposedly a minimal cost compared to the great benefits that we will obtain ... I think they really have the nerve because we all know that nowhere in the world is it possible to carry out clean extraction.

Within *destruction*, there is much story-telling about technology's inability to solve environmental issues. Based on past experiences with obsolete and outdated technology, many accounts reject the notion that technological solutions can protect future oilfields in Amazonia from environmental impacts. Hence, risk is enacted not as hypothetical consequences or minimal probability of failure due to specific associations in space and time but as a condition that people and nonhuman entities encounter and embody on a regular basis in their everyday lives. This ordering mode particularly tells about one experience, which is close to a worst-case scenario: The operations of the transnational oil company Texaco-Chevron. During the years the company operated in the Amazon, it probably caused one of the biggest environmental disasters in the history of the petroleum industry. While operating the trans-Ecuadorian pipeline (Sistema de Oleoducto Transecuatoriano; SOTE), approximately 16.8 million gallons of crude spilled from the line into the environment (Pigrau 2012: 1). Similarly, the company left hundreds of open wastewater pits without taking proper safety measures, according to environmental standards. In 1993, the local population filed a lawsuit against Texaco to force the company to accept responsibility and pay for the clean-up of the devastated areas. The case has been going on for years, and the affected communities are still waiting for a final verdict as Texaco denies any legal responsibility for the environmental damage.

Overall, *destruction* is an ordering mode that presents oil extraction in environmentally and culturally diverse territories as a highly uncertain enterprise. There are no guarantees that things will not go wrong. On the contrary, experience tells us otherwise and, more often than not, they do. Technology is far from being transparent. McNeish and Borchgrevink emphasize that with the collision of the modern scientific ideal and other knowledge systems, it "becomes more and more obvious that technological change does not remove uncertainty" (2015: 17). By telling about the effects of *poorly practiced* oil, oil's materiality is problematized, as it generates risky associations and ruptures in the Amazonian rainforest. Although there are no technological or economic heroes in this mode of ordering, there is certainly a villain among the Ecuadorian informants: Texaco-Chevron. The recurrent telling about the giant multinational oil company that violated the rights of 30,000 local inhabitants is a meaningful appropriation of the universal narrative about David versus Goliath, the powerless versus the powerful, center versus periphery, or economic power versus moral power and resistance. This is where *oil as destruction* transits towards *oil exploitation as a violation of rights*: Not only human rights but also the Rights of Nature. Consequently, the Texaco case becomes an important intersection where agency lost and agency regained meet. This movement or transition is clearly present in an informant's account regarding the lessons the environmental community has gained from the lawsuit against Texaco-Chevron[9]:

> With time, at some point in history, justice will be served. Those crimes cannot remain unpunished. I am specifically talking about Texaco. We are talking about maybe the fourth-biggest oil company on the planet, which has lost [the case] against 30,000 poor Indians and farmers in a lawsuit and even if they refuse to pay at present, at some point, they will have to do so. [Social] struggles must not decline because, at some point, they will reach an outcome. That is one of the lessons, and due to that lesson, we are still fighting for and maintaining the Yasuní issue.

The movement from ordering petroleum production in sensitive areas primarily as a matter of environmental degradation with ethical and moral implications towards telling about rights systematized within legal frameworks and international law transforms continued oil extraction into a matter of global environmental justice in several ways. First, there is the principle of "polluters pay" in a North–South perspective in which the North should pay for having largely contaminated what is often considered a global common, the atmosphere. Second, there has been a globalization or internationalization of the issue of the Amazon rainforest (Espinosa 1998: 32) due to climate change and to the role that tropical rainforests fulfill as providers of ecosystem services, the most important being carbon sinks. However, the existing disparities between countries regarding CO_2 emissions and carbon offsets have been criticized, as the current climate regime has a tendency to reproduce core–periphery dynamics between North and South (Bridge 2011: 825). Focus is placed on how "offsets and sequestration projects license the continued development and extraction of fossil fuels" (Bridge 2011: 829). Hence, this ordering mode is not only preoccupied with displaying national violations with regard to rights. On the contrary, it "goes global" as it orders a variety of linkages and connections by tying together actors and entities that are simultaneously located in many different places so as to expose the existing asymmetries related to the framing of climate change.

Moreover, the boundaries are not always clear between *destruction* and *violation of rights* as to when one stops and the next takes over since these ordering modes are often enacted together in the accounts as cause and consequence: Environmental damage calls for a framework within which this can be legally addressed. In other words, environmental issues require legal translations to become tangible offenses. Without a standardized legal framework that includes typified offenses and corresponding sanctions, there is actually no environmental crime. Laws are always performative, as they simultaneously produce legal rights and the potential violation of those same rights. When Ecuador's Constituent Assembly in 2008 included a set of articles in the supreme law known as the *Rights of Nature*, nature acquired a different *"nature"* or reality, as it went from simply being an object for human management and administration to a subject of law. The Ecuadorian Constitution, in this way, provided nature with a different kind of agency, despite practical problems related to the issue of representation (who speaks on behalf of nature). An informant who worked on these matters during the constituent process explained some of the political challenges:

> At a time when the climate crisis is evident, we have to change our relationship with nature. To stop considering nature an object and consider her a subject made a lot of sense despite the complicated legal implications and those kinds of things … How to deal with the "voice" issue, to be the legal representative of nature through a representation just like a father and a newly born or underage person, etc. It made a lot of sense. The implementation is very complicated obviously.

The difficulties and legal obstacles related to the process of implementing and practicing the Rights of Nature are also part of this ordering mode. Hence, it tells about the tension and contradiction between nature's subject status and the limited agency given to the Rights of Nature in practice. Consequently, *violation of rights* produces storylines about the potential for agency that lies in nature's constitutional rights. However, by lacking interconnectedness with secondary legislation, which has yet to be developed in a more detailed fashion, the Rights of Nature have had limited applicability so far. Hence, *violation of rights* discusses the state's ambiguous role as judge and jury, as it both facilitated the inscription of nature's rights during the constituent process and constrained their implementation at present. Consequently, the Rights of Nature are left with a limited field of action within the legal domain. They have nevertheless become important tools of contestation. Civil society groups have managed to reformulate their agency by assembling the Rights of Nature together with other strategic elements as a way to oppose and question the expansion of the oil frontier and the further extraction of fossil fuels in vulnerable areas. In the next chapter, I will discuss further how the Rights of Nature became important *allies* during the international campaign for the Yasuní-ITT Initiative and later transformed into a site of contestation by civil society groups when the government announced the decision to liquidate the project in 2013. In the following section, I will look into territorial dynamics in the Amazon where subsoil petroleum reservoirs strongly influence the spatial organization and determine relations between various actors above the surface.

The circuit of horizontal and vertical territories

The notion of territory is central to various socioenvironmental conflicts in the Amazon region. I will, therefore, start this section by looking into different ways of conceiving and experiencing territory and, more specifically, examining how these conceptions not only collide but also become elements of negotiation and resistance. While the term territory is frequently employed when referring to a delimited and formally acknowledged geographical space, it is also used when referring to more indefinite or *fluid* forms of spatial organization. Consequently, a territory clearly extends beyond the physical dimensions of space as it is produced and reproduced by a set of relations, interactions, and practices that have been performed over time and, therefore, hold the territory together. Naturally, the spatial or material aspects of territories are not necessarily fixed and invariable dimensions, that is, with clear boundaries, but they can develop and move along with relationships and their transformations. Avci and Fernández-Salvador (2016) emphasize that territory involves the symbolic and affective character of relationships towards a biophysical space. Furthermore, they emphasize that the appropriation of a territory implies limiting and controlling space and the relationships and people within that space (Avci and Fernández-Salvador 2016: 914). Over the years, the concept of territory has been fruitful for reflecting on identity politics and has, therefore, actively been used by different social movements, environmental non-governmental organizations (NGOs), and ethnic groups in the Amazon as a way to address the relationship between culture, nature, and place (Escobar 2001: 159). This way of conceiving territory frequently contradicts state policies that disrupt indigenous territorial conceptions based on their vision of life. Within an indigenous understanding, the territory is the place where life is formed, a vital space, but it is not only a space; it is also a formative process (Cisneros 2007: 134). However, different indigenous groups, or *nacionalidades* (nationalities), the political term now used in Ecuador, have different ways of experiencing and relating to their territories. When turning to the notion of territory as an analytical tool, we are confronted with a rather heterogeneous reality. When these diverse territories respond to national economic policies and capitalist insertion through the extraction of hydrocarbons, different degrees of resilience, opposition, and capacity to negotiate will influence the outcome.

Still, tensions between different conceptions and practices regarding territory are largely the result of the subsoil location of resources such as oil and minerals. These "vertical territories" influence spatial organization and the use of territories above the surface, or what we may refer to as "horizontal territories". Conflicting interests in the Amazon become evident when what indigenous communities claim as their ancestral territories are partially invaded and deforested to pave the way for petroleum activity and development. Hence, these territories are not only incorporated and assembled as economic territories within global capitalism, but their environment, their relationships, and sociomaterial practices are also reassembled in that very same process. The dynamic of spatial organization in the Amazon can be understood as the direct consequence of the interconnectedness between the underground petroleum reservoirs and the above-ground territories. This means that surface territoriality, which refers to how "social and political power are organized and exercised over space" (Brenner et al. 2003 in Bridge et al. 2013: 336), is entangled in underground territoriality. The outcome of this dynamic is a circuit of partial connections, where the Ecuadorian Amazon "is neither singular nor plural, neither one nor many" (Strathern 2004: 54).

In Ecuador, as in many countries, the Constitution establishes that the underground and its resources belong to the state. The contradiction between horizontal and vertical forms of territorialization becomes evident as the state has the right to exploit petroleum resources even if these are located within indigenous territories. A former representative of the Pachamama Foundation explained how protecting territories becomes challenging due to the institutionalization of the underground and non-renewable resources as state property:

> The biggest problem is that in Ecuador, in all our 20 constitutions, we have considered that the non-renewable resources belong to the state. That is the strongest institutional obstacle we have ... because, in the end, the state is supposedly all of us, but it is not all of us. In the end, it is the government, and if it is everyone, it depends on a national consensus, right, but that does not happen. The thing is that the indigenous peoples are there or peoples in resistance who do not want mining or oil. They want to legally protect their territory. They can protect what is above the subsoil, but there is nothing to do with the underground. So, that is the strongest institutional obstacle we have.

Hence, oil deposits become a source of power that strongly affects territorial organization in the Amazon, where overlapping and "schizophrenic" policies enable the state to carry out oil extraction in protected areas (Fontaine 2007: 355). In other words, "petroleum's matter 'territorializes'" (Valdivia 2008: 472) and, thereby, becomes a powerful actant with the capacity to modify cultural practices by changing the surrounding environment. The roads the oil companies build are one example. From the early 1970s, there has been a strong correlation between roads, colonization, and deforestation. Little (1992) explains that this has to do with companies building more roads where they have more oil wells. Colonization starts following the roads and it expands into the forest from there. Increased colonization naturally leads to higher rates of deforestation (1992: 76).

One of these roads is *vía Maxus*, built in the early 1990s by the oil company Maxus Energy Corporation. The Repsol Company operates block 16 and controls this road today, which is located within Yasuní National Park, the territory of several indigenous peoples such as the Waorani (first contacted in the 1950s) and other ethnic groups who live in voluntary isolation. By running through the rainforest, vía Maxus also runs through the social and environmental relationships of the Waorani. In 1993, the organization of the Waorani people (today *Nacionalidad Waorani del Ecuador*, NAWE) signed an agreement with Maxus that was later inherited by Repsol. According to the agreement, the oil company has to provide several services such as health, education, and community development (Rival 2015). Besides coordinating all scientific research about the Waorani and the territory, the company is also committed to giving priority to the Waorani when they need to hire workers within block 16 (Rival 2015: 267). While the presence of the oil company and the road have changed the lives of the Waorani in various ways, I will only mention a few things. As the company provides the communities with food and largely covers their basic needs, the Waorani, who have traditionally been collectors and hunters – in other words, people who have constantly been moving around – have become much more sedentary due to the presence of the oil company. Hunting patterns have changed, as their traditional spears have been replaced by guns, which has increased the pressure on some species and hunting grounds. Hunting was previously a matter of subsistence. Today, however, the road and access to transportation, often provided by the oil company, has made it possible to bring the game to market. This situation has given hunting a new monetary value. Because of this new environment, the Waorani, who are passionate hunters, dedicate most of their time to hunting since many of their other activities, such as collecting, building houses, cultivating small crops, etc., are no longer necessary (Rival 2015: 298–299). Consequently, the natural abundance of an endless forest overlaps with the abundance of the oil company, as interpreted by an Ecuadorian researcher:

They do not see how they could lack something in the future since they are people with this mentality of abundance. It is difficult to understand that you could lack something in the future. When the oil companies arrived, they saw that with the companies they had a new resource which was money, material goods that came from the outside and they were also endless. If the oil company gave them a television, and it stopped working, they could ask for another one and the company would give it to them. If the oil company at a certain point says no, I will not give you another one, they will block the road until they get a new one. If they ask for money, they will get money ... It is logical; it is another endless resource in the jungle.

While some sectors within the indigenous movement interpret this situation as blackmailing and a way to dominate the Waorani and make them dependent, others perceive this dynamic more as a form of resistance (see Rival 2015). Either way, it shows some of the complexities that unfold due to territories and people that are partially connected to the international oil market and global capitalism and partially connected to local processes of shaping and reassembling the rainforest and its meaning. Violent incidents between the Waorani and non-contacted groups (massacres in 2003 and 2013) seem to be related to increased pressure on hunting grounds, crossing paths, and overlapping territories, ambiguously articulated by an expanding oil frontier:

> Successive Ecuadorian governments have tried to implement palliative measures for the plunder, perhaps the most important have been the creation of Yasuní National Park, the concession of a territory to the Waorani, the creation of the Intangible Zone, the startup of the Plan of Precautionary Measures. Obviously, all this does not resolve the key issue: The "hidden peoples" are left without their large traditional vital space, and they have not been able to understand it, and as for today they are not willing to definitively give it up.
>
> (Cabodevilla 2013: 40)[10]

The plans and documents mentioned in this quote turned out to be important *allies* in the Yasuní-ITT Initiative, together with the *Rights of Nature*, which the final section of this chapter examines.

The inscription of nature(s) in the Ecuadorian Constitution

Ecuador approved its current Constitution in 2008 after 63.93% of the population voted for it in the referendum (López and Cubillos Celis 2009: 13). Interestingly, one of the most controversial issues during the constitutional process was that nature's status radically changed, as nature went from being an object of administrative policies and interventions to a legal subject granted constitutional rights. These rights established in several articles (art. 71–74) transformed nature into a legal subject that could be represented. This novelty caused both opposition and heated debate, especially among lawyers and constitutionalists who considered this a legal impossibility, arguing that only people can have rights and obligations (Prieto Méndez 2013: 22). Alternatively, the understanding of nature as a subject encompasses several ethical principles related to the indigenous *cosmovision*, in which humans do not occupy a privileged position in the universe but engage in a more symmetrical relationship with nature based on reciprocity and balance. According to this perspective, humans are considered part of nature and her vital cycles, not entities separated from her. Humans and nature do not belong to separate domains in indigenous ontology but are understood as relational as they enter in multiple associations with one another. According to Estermann, the foundation or first principle of reality (*arche*) in Andean philosophy is not an elementary substance or entity; on the contrary, *arche* is the relation. The relation constitutes the real "substance" in Andean thinking (Estermann 1998: 95). This perspective shares important elements with ANT that focus on how things come into being through network relations and connections. This is very different from the Aristotelian tradition in which the temporal order demands the substance or entity to first exist before it can engage in any form of relation with another entity

(Estermann 1998). Hence, Western philosophy and modernity organize reality based on dualisms, in which the relation between subject and object is exterior, as the entities remain autonomous and independent. Similarly, the Andean worldview also rests on a dualist understanding of reality (feminine/masculine, micro-/macro-cosmos, up/down, moon/sun, left/right, etc.), but the components far from being independent, establish a variety of relations through reciprocity, correspondence, and complementarity (Kaarhus 1989; Estermann 1998). Hence, the inclusion of the *Rights of Nature* is a significant political achievement, as it represents, on the one hand, a formal recognition of the ancestral indigenous knowledge, and on the other hand, a first step away from an entirely anthropocentric worldview in legal matters towards a more biocentric one.

Just as shown in the laboratory studies associated with STS in the 1980s, it is possible to take a laboratory approach towards the inscription and produc-tion of new legal realities that resulted from the work of Ecuador's Constituent Assembly. I will draw on the conception of translation to explain how this political institution transformed the indigenous nature into a legal instrument for environmental governance and policy-making. Through descriptions and ordering, nature's identity and interactions were defined (Callon 1995), which brought into play a new legal and political actant. The next section looks into this process.

Opening the "black box" of nature

The Ecuadorian Constitution starts with the following preamble:

We women and men, the sovereign people of Ecuador
RECOGNIZING our age-old roots, wrought by women and men from various
    peoples,
CELEBRATING nature, the Pacha Mama (Mother Earth), of which we are a
    part and which is vital to our existence,
INVOKING the name of God and recognizing our diverse forms of religion and
    spirituality,
CALLING UPON the wisdom of all the cultures that enrich us as a society,
AS HEIRS to social liberation struggles against all forms of domination and
    colonialism
AND with a profound commitment to the present and to the future,
Herby decide to build
A new form of public coexistence, in diversity and in harmony with nature, to
    achieve the good way of living, the sumak kawsay;
A society that respects, in all its dimensions, the dignity of individuals and
    community groups;
A democratic country, committed to Latin American integration – the dream of
    Simón Bolívar and Eloy Alfaro – peace and solidarity with all peoples of the
    Earth;

And, exercising our sovereign powers, in Ciudad Alfaro, Montecristi, province of
    Manabí, we bestow upon ourselves the present:
Constitution of the Republic of Ecuador.

(Constitution of the Republic of Ecuador 2008)

As we can see from this excerpt of the text, nature is mentioned at the very beginning of the Constitution. By referring to nature as *Pachamama*, it becomes clear that
it is not an anonymous or random nature that is inscribed but the feminine and
divine nature that the indigenous communities in the Andes experience through
tradition, rituals, and everyday interaction. Behind this particular conception of
nature, we find a practice that is based on empirical knowledge gained through
lived experience and a relational understanding of nature and humans. *Pachamama*
is a living organic being (Estermann 1998: 176), and humans communicate with
her by cultivating the land and through a variety of rituals and ceremonies. More
than rational entities or producers transforming nature, human beings are understood as natural entities (Estermann 1998: 174). The indigenous communities do
not inhabit a culture that is separated from nature, as nature was always a *nature-
culture* (Latour 1993) in the indigenous communities, where humans to a large
extent become what they are through multiple relations both with other humans
and with the natural and material world.

Unlike Western thinking that emphasizes a marked distinction between subject and object, the Indians do not consider themselves subjects in a traditional
Western sense; rather, they define themselves from a relational perspective, that
is, how he or she relates to others. Humans are, therefore, not the center or origin
of all activities and action, and the human consciousness is not necessarily the
point of departure for knowledge about the world. On the contrary, human beings
are primarily understood as collective and relational subjects. As interpreted by
Estermann, the Indian in the Andes is, first and foremost, a *chakana* (bridge or a
knot) made up of multiple relations or connections. He or she becomes himself or
herself through others or, more specifically, through the relations that he or she
establishes with others (Estermann 1998: 201–202). In the Andes, human beings
are relational both regarding other human beings within society or a collective, as
well as with nature and the cosmos.

Moreover, the idea of nature as something exterior and completely different
from human beings, in short, an inanimate and brute reality (Estermann 1998:
172) goes all the way back to the Scientific Revolution when many philosophers
thought of nature as something similar to a mechanical construction or a big
machine. Boyle and Descartes, for example, considered the clock metaphor "a
philosophically legitimate way of understanding how the natural world was put
together and how it functioned" (Shapin 1996: 34). However, the main reason
why the clock metaphor became so popular was principally that the clock was an
artifact produced by a human agent to fulfill a purpose, but it was not an intelli-
gent creature with its own will or intentions (Shapin 1996). Hence, the mechanical
philosophy in general and the clock metaphor, in particular, played an important

role in the process of demystifying nature. Furthermore, only the creator, not the creation, is divine in the Judeo-Christian tradition. Since nature was created, just like the clock by the clockmaker, it was by definition profane.

The Rights of Nature

Chapter seven, art. 71, in the Constitution under the title "Rights of Nature", establishes "Nature, or Pacha Mama, where life is reproduced and occurs, has the right to integral respect for its existence and for the maintenance and regeneration of its life cycles, structure, functions and evolutionary processes". The same article also states that "All persons, communities, peoples and nations can call upon public authorities to enforce the rights of nature". Whereas traditional conceptions objectify nature by developing administrative policies and governance systems, the Ecuadorian Constitution inscribes and translates nature into a subject of law with a specific cultural identity. By using nature's/earth's Andean name, *Pachamama*, as a synonym followed by a characterization ("where life is reproduced and occurs"), nature steps out of the category of an abstract entity and becomes individualized in a certain way. Of all the possible natures, *Pachamama*, with her connotations from Andean collective imaginary, is the one that the Constitution inscribes. The Ecuadorian Constitution opens the black box of nature by doing this: There has never been a single and coherent nature but a variety of natures, and these have coexisted in periods of transition. The inclusion of *Pachamama* in the Ecuadorian Constitution questions the established Western view of nature. When nature is not considered a coherent whole anymore, something happens: The black box is reopened with new possibilities of negotiation and appropriation. This is, therefore, a way to decolonize the Constitution and the legal system. Bringing in the indigenous nature exposes the relationship between modernity and colonialism. The colonial power was not only a military and political power, but it also implied the authority to reject other knowledge systems and question their objectivity and utility. Western scientific knowledge became an ideal against which all other knowledge systems were "measured" and found uncivilized, exotic, and naïve. The idea of everything nonhuman as an instrumental and economic reality separated from the human sphere was violently imposed on the "New World". According to Latour, the modern project is based, to a large extent, on an artificial dichotomization of nature and nonhumans on the one hand and culture and humans on the other. This is what he calls the work of purification (Latour 1993). Nevertheless, the indigenous traditional knowledge system continued to exist side by side with the Western dominant scientific ideal. The performative action of granting *Pachamama* subject status and rights of her own is also a political act of emancipation, as it shows that the New World was actually never new but an ancient world with valid, accumulated knowledge and practices based on experience and reciprocity with nature. Furthermore, the Rights of Nature is also a way to question the Western instrumental understanding of nature and natural resources as unlimited, which has led to the global environmental crisis we face today. *Pachamama*, from this perspective, is not only a divine nature but also a political nature that enacts new sociopolitical realities and representations.

While constitutional articles with a biocentric approach (art. 71–73) can be seen as conflicting with those that are essentially anthropocentric, this also illustrates that political attempts to design more balanced collective relations between human and nonhumans, what Latour (2004) calls *cosmopolitics* (borrowing a term from Stengers), often require syncretic ordering modes (Law et al. 2013). The outcome of this syncretism is coexistence between several natures in the Ecuadorian supreme law. Subject nature/*Pachamama* interacts with objectified nature from the Western philosophical tradition, which can be observed in several articles. One example is art. 74: "Persons, communities, peoples, and nations shall have the right to benefit from the environment and the natural wealth enabling them to enjoy the good way of living" (Constitution of the Republic of Ecuador 2008). While this article clearly enacts nature as natural resources, many sectors, particularly environmentalists and part of the indigenous movement (not all the organizations agree), claim that art. 71–73, among others, make it difficult, if not impossible, to continue expanding the oil frontier in the Amazon without neglecting the intention of the Rights of Nature.

Notes

1 EP Petroecuador: 40 años construyendo el desarrollo del país. Informe estadístico (1972–2012), pp. 24–25.
2 API gravity compares the weight of oil in relation to water.
3 The EIA classifies crude oil with less than 1% sulfur as sweet and crude oil with more than 1% sulfur as sour (EIA 2017).
4 Both the *Oriente* and the *Napo* are sour crudes with a sulfur content of 1.45% and 2.10%, respectively (El Universo 2015).
5 EP Petroecuador: 40 años construyendo el desarrollo del país. Informe estadístico (1972–2012), p. 75.
6 Petroleum.co.uk: Heavy crude oil. http://www.petroleum.co.uk/heavy-crude-oil.
7 Rafael Correa Delgado, Ecuador's president from 2007 to 2017, was part of the previous progressive period in Latin America known as the "pink tide".
8 The millennium schools were considered part of the United Nations Millennium Goals. They have primarily been built in rural and peripheral areas where the population in the past has been excluded from quality education. The idea was to give access to education in modern buildings equipped with technology and functional infrastructure to excluded groups in their own communities. Similarly, millennium communities or "cities" have been built in areas that have suffered the consequences of oil exploitation in the Amazon as a way to distribute the oil rents in this part of the country. These projects, however, have been strongly criticized by several civil society groups and environmental NGOs that regard the projects as opposed to Amazonian communities' true needs and practices.
9 In February 2011, the Provincial Court of Justice in Sucumbíos (Ecuador) ruled in favor of the claimants, ordering Chevron to pay more than 8.6 billion dollars for the clean-up and reparation for the environmental harm (Pigrau 2012: 6). As of today, the company refuses to pay and denies all legal responsibility for the environmental damages.
10 Translation by the author.

References

Acosta, A. (1990). *La Deuda Eterna*. Quito: Libresa.

Acosta, A. (1995). *Breve Historia Económica del Ecuador*. Quito: Corporación Editora Nacional.

Acosta, A. (2009). *La Maldición de la Abundancia*. Quito: Abya-Yala.

Arsel, M., Hogenboom, B., & Pellegrini, L. (2016). The extractive imperative in Latin America. *Extractive Industries and Society*, 3(4), 880–887.

Avci, D., & Fernández-Salvador, C. (2016). Territorial dynamics and local resistance: Two mining conflicts in Ecuador compared. *Extractive Industries and Society*, 3(4), 912–921.

Blok, A., & Jensen, T. E. (2011). *Bruno Latour: Hybrid Thoughts in a Hybrid World*. London: Routledge.

Bridge, G. (2011). Resource geographies I: Making carbon economies, old and new. *Progress in Human Geography*, 35(6), 820–834.

Bridge, G., Bouzarovski, S., Bradshaw, M., & Eyre, N. (2013). Geographies of energy transition: Space, place and the low-carbon economy. *Energy Policy*, 53, 331–340.

Cabodevilla, M. Á. (2013). La masacre … qué nunca existió? In: M. Marchi, M. Aguirre, & M. Á. Cabodevilla, *Una Tragedia Ocultada* (pp. 21–139). Quito: CICAME. Fundación Alejandro Labaka.

Callon, M. (1995). Four models for the dynamics of science. In: S. Jasanoff, G. E. Markle, J. C. Petersen, & T. Pinch, *Handbook for Science and Technology Studies* (pp. 29–63). Thousand Oaks: Sage.

Callon, M. (1999). Actor-network theory: The market test. *Sociological Review*, 47(1), 181–195.

Cisneros, P. (2007). Los conflictos territoriales y los límites de la cogestión ambiental. In: G. Fontaine, & I. Narváez, *Yasuní en el Siglo XXI: El Estado Ecuatoriano y la Conservación de la Amazonía* (pp. 129–174). Quito: Flacso.

Constitution of the Republic of Ecuador. (2008, October 20). Retrieved from Political Database of the Americas: http://pdba.georgetown.edu/Constitutions/Ecuador/engl ish08.html.

Davis, G. A. (1995). Learning to love the Dutch disease: Evidence from the mineral economies. *World Development*, 23(10), 1765–1779.

Davis, G. A., & Tilton, J. E. (2005). The resource curse. *Natural Resource Forum*, 29(3), 233–242.

EIA. – U.S. Energy Information Administration (2017, September 21). *Changing Quality Mix Is Affecting Crude Oil Price Differentials and Refining Decisions*. Retrieved from https://www. eia.gov/todayinenergy/detail.php?id=33012.

EIA. – U.S. Energy Information Administration (n.d.). *Petroleum & other liquids. Definitions*. Retrieved from http://www.eia.gov/dnav/pet/TblDefs/pet_pri_wco_tbldef2.asp.

El Universo. (2015, January 23). *Castigo que recibe petróleo de Ecuador es menor en enero*. Retrieved from http://www.eluniverso.com/noticias/2015/01/23/nota/4468431/castigo-que-recibe-crudo-ecuador-es-menor-enero.

Endara G. (ed). (2014). *Post-Crecimiento y Buen Vivir: Propuestas Globales para la Construcción de Sociedades Equitativas y Sustentables*. Quito: Friedrich Ebert Stiftung, ILDIS.

EP Petroecuador. (n.d.). *40 Años construyendo el desarrollo del país 1972–2012. Informe estadístico*. Retrieved from http://eppintranet.eppetroecuador.ec/idc/groups/public/docum ents/peh_docsusogeneral/ep003014.pdf.

Escobar, A. (2001). Culture sits in places: Reflections on globalism and subaltern strategies of localization. *Political Geography*, 20(2), 139–174.

Espinosa, M. F. (1998). La Amazonía ecuatoriana: Colonia interna. *Íconos: Revista de Ciencias Sociales*, 5(5), 28–34.

Estermann, J. (1998). *Filosofía Andina: Estudio Intercultural de la Sabiduría Autóctona Andina.* Quito: Ediciones Abya-Yala.

Fontaine, G. (2002). Sobre bonanzas y dependencia: Petróleo y enfermedad holandesa en el Ecuador. *Íconos: Revista de Ciencias Sociales*, 13(13), 102–110.

Fontaine, G. (2007). *El Precio del Petróleo: Conflictos Socio-Ambientales y Gobernabilidad en la Región Amazónica.* Quito: Flacso, IFEA, Abya-Yala.

Grosfoguel, R. (2000). Developmentalism, modernity, and dependency theory in Latin America. *Nepantla: Views from the South*, 1(2), 347–374.

Gudynas, E. (2005). América Latina: Geografías fragmentadas: Sitios globalizados, áreas relegadas. *Revista del Sur*, 3–13.

Hurtado, O. (1997). *El Poder Político en el Ecuador.* Quito: Planeta-Letraviva.

Kaarhus, R. (1989). *Historias en el Tiempo, Historias en el Espacio. Dualismo en la Cultura y Lengua Quechua/Quichua.* Quito: Ediciones Tinkui, Abya-Yala.

Karl, T. L. (1997). *The Paradox of Plenty: Oil Booms and Petro-States.* Berkeley: University of California Press.

Karl, T. L. (2007). Oil-led development: Social, political, and economic consequences. *CDDRL Working Papers, Freeman Spogli Institute for International Studies*, 80, 1–36.

Latour, B. (1987). *Science in Action: How to Follow Scientists and Engineers Through Society.* Cambridge, MA: Harvard University Press.

Latour, B. (1993). *We Have Never Been Modern.* Cambridge, MA: Harvard University Press.

Latour, B. (2004). *Politics of Nature: How to Bring the Sciences into Democracy.* Cambridge, MA: Harvard University Press.

Latour, B. (2005). *Reassembling the Social: An Introduction to Actor-Network Theory.* New York: Oxford University Press.

Law, J. (1994). *Organizing Modernity.* Oxford: Blackwell.

Law, J., & Singleton, V. (2014). ANT, multiplicity and policy. *Critical Policy Studies*, 8(4), 379–396.

Law, J., Afdal, G., Asdal, K., Lin, W.-y., Moser, I., & Singleton, V. (2013). Modes of syncretism: Notes on noncoherence. *Common Knowledge*, 20(1), 172–192.

Little, P. E. (1992). *Ecología Política de Cuyabeno: El Desarrollo No Sostenible de la Amazonía.* Quito: ILDIS, Abya-Yala.

López, A. A., & Cubillos Celis, P. (2009). Análisis del referendum Constitucional 2008 en Ecuador. *Íconos: Revista de Ciencias Sociales*, 33(33), 13–20.

McNeish, J. A., & Borchgrevink, A. (2015). Introduction: Recovering power from energy: Reconsidering the linkages between energy and development. In: J. A. McNeish, A. Borchgrevink, & O. Logan (eds), *Contested Powers: The Politics of Energy and Development in Latin America* (pp. 1–39). London: Zed Books.

Mol, A. (2002). *The Body Multiple: Ontology in Medical Practice.* Durham: Duke University Press.

Mol, A., & Law, J. (2002). Complexities: An introduction. In: J. Law, & A. Mol (eds), *Complexities: Social Studies of Knowledge Practices* (pp. 1–22). Durham: Duke University Press.

Müller, M. (2015). Assemblages and actor-networks: Rethinking socio-material power, politics and space. *Geography Compass*, 9(1), 27–41.

North, L. L., & Grinspun, R. (2016). Neo-extractivism and the new Latin American developmentalism: The missing piece of rural transformation. *Third World Quarterly*, 37(8), 1483–1504.

Oil price.com. (2009, December 2). *A detailed guide on the many different types of crude oil.* Retrieved from http://oilprice.com: http://oilprice.com/Energy/Crude-Oil/A-Detail ed-Guide-On-The-Many-Different-Types-Of-Crude-Oil.html.

Petroleum.co.uk. (n.d.). *Heavy Crude Oil.* Retrieved from www.petroleum.co.uk: http:// www.petroleum.co.uk/heavy-crude-oil.

Pigrau, A. (2012). *The Texaco-Chevron case in Ecuador. EJOLT Factsheet.* Retrieved from www. ejolt.org: http://www.ejolt.org/wordpress/wp-content/uploads/2015/08/FS-42.pdf.

Prieto Méndez, J. M. (2013). *Derechos de la Naturaleza: Fundamento, Contenido y Exigibilidad Jurisdiccional.* Quito: Corte constitucional del Ecuador. Centro de Estudios y Difusión del Derecho Constitucional (CEDEC).

Quevedo, C. E. (1986). El sector energético ecuatoriano y la caída de los precios internacionales de petróleo. In: S. Escobar (ed). *Ecuador: Petróleo y Crisis Económica* (pp. 91–150). Quito: ILDIS.

Rival, L. M. (2015). *Transformaciones Huaoranis: Frontera, Cultura y Tensión.* Quito: Universidad Andina Simón Bolívar, Latin American Centre of Oxford, Abya-Yala.

Robertson, R. (1995). Glocalization: Time-space and homogeneity-heterogeneity. In: M. Featherstone, S. Lash, & R. Robertson (eds), *Global Modernities* (pp. 25–44). London: Sage.

Ross, M. L. (1999). The political economy of the resource curse. *World Politics*, 51(2), 297–322.

Sachs, J. D., & Warner, A. M. (1995). Natural resource abundance and economic growth. *NBER National Bureau of Economic Research Working Paper No. 5398* (pp.1–47).

Sachs, J. D., & Warner, A. M. (2001). The curse of natural resources. *European Economic Review*, 45(4–6), 827–838.

Shapin, S. (1996). *The Scientific Revolution.* Chicago: University of Chicago Press.

Stevens, P. (2003). Resource impact: Curse or blessing? A literature survey. *Journal of Energy Literature*, 9(1), 3–42.

Strathern, M. (2004). *Partial Connections.* Walnut Creek: Roman & Littlefield.

Valdivia, G. (2008). Governing relations between people and things: Citizenship, territory, and the political economy of petroleum in Ecuador. *Political Geography*, 27(4), 456–477.

Valdivia, G. (2015). Oil frictions and the subterranean geopolitics of energy regionalisms. *Environment and Planning A: Economy and Space*, 47(7), 1422–1439.

Wallerstein, I. (2004). *World-Systems Analysis: An Introduction.* Durham: Duke University Press.

5   Becoming Yasuní-ITT

A process of assembling and reassembling

This chapter addresses the process of the Yasuní-ITT (Ishpingo, Tambococha, and Tiputini) Initiative as an evolving design that assembled and enrolled a series of elements and actors in an attempt to go global. I will use the pairing concepts of framing and overflowing (Callon 1998a) to examine how the Yasuní Initiative *overflowed* the United Nations Framework Convention on Climate Change (UNFCCC) and the Kyoto Protocol as it tried to open the black box of climate change mitigation. Emphasis is therefore placed on tracing some of the documents and the different parallel processes that were assembled in the Yasuní-ITT Initiative. This chapter also analyzes how the struggle to keep the oil in the ground was translated as a matter of defending the rights of humans and nonhumans. Finally, the chapter addresses the role of civil society in reassembling the initiative as a site of contestation and democratic practice.

Although Yasuní-ITT is a place-based controversy, it would be misleading to conceptualize it as place-bound, as it is intersected by actors and actions found at different levels or scales. Regarding this last aspect, it is important to keep in mind "that scale does not depend on absolute size but on the number and qualities of dispatchers and articulators" (Latour 2005: 196). Consequently, the political processes and debates evolving around the ITT field extend beyond the local geographic area, as a series of frameworks, documents, policies, and scientific research links local challenges related to petroleum extraction and environmental degradation to global climate change mitigation. In practice, the formalization of the Yasuní-ITT Initiative as state policy can be understood as an attempt to stabilize and make durable a local/national network by enrolling international actors and entities: In other words, to give it more and stronger connections and interrelatedness. Esperanza Martínez, an Ecuadorian environmentalist and activist, has described the emergence of the initiative in the following terms:

> The proposal of not exploiting the crude from the fields of Ishpingo-Tambococha-Tiputini (ITT) in Yasuní National Park in Ecuador had to laboriously make its way at the local, national, and international levels. Consequently, from Kyoto to Quito is more than only a metaphor, it is a long journey, that attempts to disassemble visions about climate change, about the development model, and about the rights of peoples and nature.[1]
>
> (2009: 7)

The quote suggests that the idea of leaving the petroleum reserves underground was shaped over time and across space by a series of events, entities, and actors as the proposal went through several translations in an expanding and changing network. Specifically, we are referring to "an on-going process made up of uncertain, fragile, controversial, and ever-shifting ties" (Latour 2005: 28). As precarious achievements, network translations do not necessarily lead to successful outcomes. This is largely the case of the Yasuní Initiative, which is frequently cataloged as a policy failure despite receiving considerable national and international attention and support. Nevertheless, it constitutes a relevant example of challenges related to a possible transition towards a different kind of development model less dependent on oil resources. In other words, the initiative had the potential of "allowing a transition towards a post-petroleum society" (Larrea and Warnars 2009: 219). However, one of the main issues that caused ambiguity is the fact that Ecuador has a highly oil-dependent economy, and revenues from extractivism are "still considered to be the basis of Ecuador's emerging development strategy" (Gallardo Fierro 2016: 939). Consequently, Yasuní, as part of an expanding oil frontier, represents a common dilemma, namely, how to solve the tension between conservation of nature and development (Arsel 2012; Pellegrini et al. 2014; Gallardo Fierro 2016). By seeking compensation from the international community for not extracting the crude from the ITT field, the Ecuadorian government tried to solve the conflict in a rather holistic manner by presenting a proposal that took into consideration local, national, and international issues.

Although the initiative did not manage to maintain sustained adherence by key international participants or contributors over time, it constitutes an interesting case of conflicting interests and logics that led to ambiguous enrollment strategies and diffuse alliances. Hence, the Yasuní-ITT case can be analyzed from an actor-network theory (ANT) perspective, as several moments of translation produced a dynamic that later prevented institutionalization and positioning of the proposal as an international mechanism not only to protect biological and cultural diversity but also to address the global climate crisis. Studying both successful outcomes and failures has a long tradition within science and technology studies (STS) and dates back to Bloor's Strong Programme and the principle of symmetry. Focus is placed, therefore, on how specific actors and actants are invited, enrolled, or excluded from participating in the network and how agendas set by significant actors enable, transform, or block connections or linkages (Bartlett and Vavrus 2017: 45). Overall, the intention of designing a new mechanism, which was not directly anchored in the Kyoto Protocol, was probably one of the initiative's main difficulties. For technocrats who work with project development within the field of climate change, it was clearly a disadvantage. As stated by a former member of the Administrative Committee for the Yasuní-ITT Initiative, a recurrent argument was that "it does not exist in the Kyoto Protocol so we cannot do it". Conversely, an informant from an Ecuadorian environmental organization believed "the Yasuní-ITT Initiative was a way to step out of the box and reject Kyoto and the carbon markets". As the Yasuní-ITT Initiative developed over time due to national and international discussions and requirements, it ended up including new elements, thereby also articulating a somewhat different focus. In its final version, it contained a logic based on carbon budgeting (Sovacool and Scarpaci 2016: 159), which implies that the majority of the world's fossil-fuel reserves constitute unburnable carbon and must be kept in the ground. One way to respond to climate science is the current UNFCCC scheme based on specific targets for reducing carbon dioxide (CO_2) emissions. The other part is still missing, namely an international agreement on a mechanism to forego fossil-fuel reserves and, more specifically, where and how these should be kept permanently locked in. Hence, the Yasuní-ITT Initiative comprised an important contribution to a still incipient debate by providing a formula: The world should start stranding petroleum assets in developing countries that are classified as megadiverse and "located between the tropics of Cancer and Capricorn, where tropical forests are concentrated" (Larrea et al. 2009: 5). Against this background, despite not achieving the envisioned outcome, the Yasuní Initiative "provides rich insights into the types of challenges that arise when trying to strand crude oil assets in practice" (Sovacool and Scarpaci 2016: 159). In the next section, I will specifically look into some of the challenges related to the process of designing non-extraction policies for the oil deposits in the ITT block. The complexity of the process demands a "glocal" lens, as the political challenges transcend the socioenvironmental conflicts in the Amazon through several translations where the climate issue becomes "localized" in Yasuní's oil reserves, while the local forest is "globalized" as carbon sinks and, thereby, becomes an important atmospheric strategy.

Political challenges and an evolving design

What originally started as a civil society demand for an oil moratorium in the Ecuadorian Amazon transformed as it went through different phases and encountered several political challenges before reaching the status of a national policy. Several decades of petroleum production had clearly taken its toll on the environment and, thereby, severely affected the local population. Hence, the processes surrounding the Yasuní-ITT Initiative can be traced back to local struggles and demands put forward by indigenous communities and environmental organizations back in the early 1990s. After experiencing the consequences of petroleum exploitation in the northern Amazon, local actors started to denounce both the negative impacts on fragile ecosystems caused by the oil industry and the severe degradation of the rainforest that was affecting their livelihoods. Consequently, the resistance and questioning of the legitimacy of the oil industry and the way it carried out its operations began to gain adherence not only locally but also nationally among different civil society groups and organizations. This resistance was largely, but not exclusively, related to Texaco's operations between 1964 and 1992 that contaminated soil, rivers, and waterways with toxic substances and oil spills. Besides leaving behind hundreds of open wastewater pits without carrying out any kind of environmental procedures to clean up the area, the company's operations are also associated with considerable deforestation. Overall, "the deforestation of 2,000,000 hectares of land is attributed to petroleum operations in the northern Ecuadorian Amazon" (Pigrau 2012: 1). The social impacts are another type of risk generated by asymmetric power relations between indigenous communities and the oil companies, wherein the local population ends up depending on the company to obtain services or benefits as part of so-called friendly agreements and community relations that allow the oil companies to gain access and operate within indigenous territories. These agreements have produced social fragmentation and divided the population on several occasions, as the negotiations with the oil companies often provide only temporary benefits and employment for the community at the expense of the environment and healthy ecosystems. In other words, the negotiations and relationship with the oil companies often imply a process in which local grassroots organizations merely play an instrumental role (Fontaine 2007a: 279), in line with the oil companies' objectives.

Over time, socioenvironmental conflicts and externalities became important elements assembled together in processes of local resistance and territorial defense. Nationally based non-governmental organizations (NGOs), together with transnational environmental organizations and networks, have also been important actors mobilizing against a further expansion of the oil frontier towards the central-southern part of the Amazon region. According to Fontaine (2007a), in Ecuador, the environmental organizations with an activist profile can be roughly divided into two categories: On the one side are those that take a moderate stance towards the presence of the oil companies and emphasize controlled extraction with environmental standards. In this case, the main focus tends to be on the local communities' right to self-determination, and their work is often oriented towards informing people about their rights besides providing them with the necessary instruments that will enable them to negotiate on more equal terms with the oil companies. On the other side, we find the radical organizations that generally oppose all forms of extractive industries and, therefore, try to mobilize the communities against the oil companies to resist their incursion into local territories (Fontaine 2007a: 289–290). However, on several occasions, these organizations have joined their efforts; for example, during the campaign *Amazonía por la vida* (Amazon for life), initiated in 1989, several NGOs decided to coordinate their activities to articulate various existing initiatives with the purpose of better defending the Amazon region and the local population (Acción Ecológica 2000). Environmental organizations had repeatedly presented proposals in the past to demand a moratorium on further petroleum extraction in Amazonia. The Pachamama Foundation, the Center for Economic and Social Rights/Centro de Derechos Económicos y Sociales (CDES), and Acción Ecológica presented an initiative to the Ministry of the Environment in 2003. Some years later, in 2007, the Pachamama Foundation proposed a *green plan* to pursue sustainable development in the central-south of the Amazon (Martínez 2009: 17). The plan requested an oil moratorium as a means to guarantee the conservation and sustainable management of biodiversity by rethinking the current development model in this part of the country. The plan proposed an agreement for institutional cooperation between Pachamama, the Ministry of the Environment, and the Ministry of Economy and Finance to develop technical input that could facilitate decision-making at the level of these ministries (Fundación Pachamama 2007). Additionally, the plan contemplated the possibility of reorienting mechanisms such as debt swaps and accessing new financial sources available due to global climate change (Fundación Pachamama 2007). This background contains significant elements and ideas that later were developed and translated into the Yasuní Initiative. The demand for an oil moratorium as a necessary requirement to defend indigenous territories and protect biodiversity was linked over time to climate change mitigation. Consequently, the climate issue was enrolled as an important *ally* that was worked into the designs of potential conservation mechanisms. As an international network, climate change and, more specifically, the international climate regime operates as an actor-network with the ability to bridge local and global challenges by connecting networks and resources. A representative from Acción Ecológica, one of the more radical environmental organizations, explained the idea of an oil moratorium in the Amazon and its links to the global climate issue as follows:

> The idea of an oil moratorium is an old idea of many organizations, primarily environmental [organizations] and also many indigenous organizations but maybe a bit more verbalized and positioned at the international level by organizations working with environmental issues. It was an idea that was constructed around the protection of territories, on some occasions, natural areas, national parks, and on other occasions, indigenous territories. As the climate issue becomes increasingly important in international discussions, the moratorium actually started with avoiding climate change as one of its arguments, the moratorium is increasingly positioned as a real alternative to avoid climate disasters ... So, the discourse within society originated as a matter of territorial defense and is positioned at the international level as an alternative to avoid climate issues, and as an alternative to avoid the climate issue from continuing down the same path where it started, because it was quickly co-opted by market logics.

The thesis of an oil moratorium is clearly presented in the informant's account as the element that brings territorial struggles and global climate policies together. While the thesis represents an important background for the ITT Initiative, the government's proposal was even more radical as it took the idea a step further. A moratorium implies a temporary prohibition or the action of postponing something, in this case, an expansion of the oil frontier. Alternatively, the declared intention of the Yasuní-ITT Initiative was to keep the oil underground permanently. Hence, to understand the various political, cultural, and economic challenges tied to the proposal of stranding petroleum assets in Yasuní, we need to take a closer look at the process of assembling the territory and how different objectives, underpinned by diverse logics, have produced policies of conflicting spatial organization. It is important to keep in mind that an assemblage constitutes a mode of ordering that, for a certain time, holds together heterogeneous entities, which constitute a new whole (Müller 2015: 28). Consequently, Yasuní National Park can be described as a territory that was always "in the making" and characterized by shifting territorial extensions and variations in its borders and zoning schemes. Politically, it was always an uncertain, problematic, and unsettled territory, as it never really comprised a single or coherent place but several environmental, economic, and cultural territorial projects in tension. The outcome can best be understood by turning once more to the concept of partial connections. As I explained in the previous chapter regarding the Amazon, Yasuní is distributed among several actors, legal frameworks, and policy areas that are found simultaneously at the local, national, and global levels. As a complex circuit of connections, Yasuní "cannot be approached holistically or atomistically, as an entity or as a multiplication of entities" (Strathern 2004: 54) as it largely constitutes *a territory of "in-betweens"*. The next section examines how these in-betweens are made to cohere despite being *unfitting*, as they clearly constitute political disparities.

The spatial organization of conflicting territorialities

As the previous chapter discussed, the underground has a preponderant role in past and present processes of spatial organization in the Ecuadorian Amazon. Yasuní is a clear example in this context of how vertical territories in practice influence the distribution or allocation of space above the surface. In other words, the oil-bearing formations located in the underground constitute an important point of departure for the various processes of territorialization, which include assembling Yasuní as an oil-producing space but also as a location where the state aims to reconcile multiple purposes and conflicting agendas. These circumstances have led to shifting government policies translated into maps that assemble and reassemble the territory's extension and, thereby, its spatial dynamics. It is important, therefore, to take into account that "resource-making activities are fundamentally matters of territorialization" (Bridge 2011: 825). Yasuní National Park was created in 1979 with a total extension of 678,000 hectares that was later reduced to 544,730 hectares in 1990. However, two years later, the park's borders were once more modified when the government extended the territory to include a total of 982,000 hectares of forest (Fontaine and Narváez 2007; Álvarez 2013). The motivation behind this continuous process of re-designing or re-making Yasuní as a territory is primarily tied to extractive interests and national dependence on oil revenues (Álvarez 2013: 81). Figure 5.1 shows these shifting territorializations of Yasuní National Park, where the three maps display the limits from 1979, 1990, and 1992. The underground territories are significant *actants* that modify the geographic and relational landscape of Yasuní. In the foregoing chapter, I described how the construction of roads and transportation systems altered the relationship between the forest, humans, animals, and ancestral territories and how different territorial dynamics directly affect important ecosystems. Although the roads that follow the pipelines are an essential part of the problem, deforestation actually begins with geological prospecting when preparing seismic survey lines. When opening trails for exploration, it is not uncommon to deforest hundreds of kilometers within an area of 500,000 hectares (Fontaine 2007a: 277–278). Landing strips for airplanes and helicopters and the use of explosives to determine the

Límites a 1979

Límites a 1990

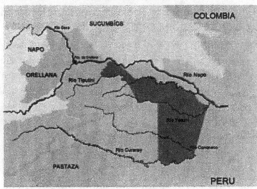

Límites a 1992

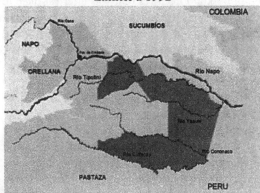

Figure 5.1 The legal limits of Yasuní National Park (1979, 1990, and 1992). Produced by Paúl Cisneros in Fontaine and Narváez, 2007. Source: IGM, 2006; ECORAE, 2003.

location of petroleum deposits also disturb various species whose habitats are where these activities occur (Fontaine 2007a: 278). Deforestation later continues around exploration wells as oil workers set up their camps in the forest by using timber from the area. If some of the oil wells turn out to be profitable, they will require the necessary infrastructure: Platforms, pipelines, access roads, and additional camps for the oil workers. These are all circumstances that attract further human settlement, land use, and commercial activity with local species and bushmeat, increasing the pressure on wildlife populations. These are just a few examples of how the underground and vertical territorializations influence, and even accommodate, the ordering of space, nature, and humans above the surface. It is important, therefore, to think of "territoriality as a fundamentally relational and processual concept with a strong agency element" (Larsen 2017: 51).

We find multiple territorialities at work in Yasuní National Park, and many of them overlap and compete for space. The park is, therefore, not only a territory that has been moved around by several ministerial agreements and decrees as the maps illustrate, but it is also a territory with several conflicting territories within. Today, there are six oil blocks (14, 15, 16, 17, 31, and 43) that are located completely within the park's boundaries or that partially overlap with the park or the buffer zone. The territorial overlaps between oilfields and protected areas comprise part of the state's attempt to coordinate economic and conservationist aims within the same area. The Management Plan for Yasuní National Park, issued by the Ministry of the Environment, states its first objective in chapter 4: "To manage protection and sustainable use of natural resources in Yasuní National Park and its buffer zone"[2] (Ministerio del Ambiente 2011: 26).

However, Ecuador's dependence on oil export revenues has frequently made environmental governance a secondary matter, a situation reflected in a zoning scheme that gives the petroleum industry access and guarantees oil production in specific areas in the Amazon region (Narváez et al. 2013: 9). In Yasuní, oil blocks overlap not only with protected areas that are environmentally vulnerable but also with the Intangible Zone (Zona Intangible Tagaeri Taromenane; ZITT) created to protect uncontacted indigenous groups and the Waorani territory. However, the latter is excluded from the petroleum map produced by the Ministry of Energy and Non-Renewable Natural Resources (see Figure 5.2) and, thereby, rendered invisible, as the map enacts the state's assembling of Yasuní primarily as a twofold enterprise in which environmental governance is juxtaposed with industrial production. By leaving other coexisting zones and indigenous territorialities unmapped, Yasuní is enacted as a less problematic and less uncertain territory than is actually the case. Since conflicting visual elements are kept to a minimum (national system of protected areas and the ZITT), the petroleum map features a condensed version of spatial conflicts in which other complicating issues are folded up and smoothed out.

Hence, the map depicts the Amazon region principally as an extractive territory where oil blocks appear as layered on top of the national park and the rest of the Ecuadorian Amazon, where there seems to be nothing except for virtual

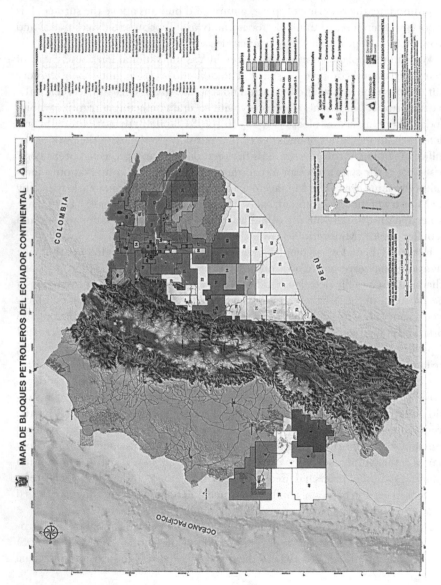

Figure 5.2 Map of oil blocks in continental Ecuador. Source: Secretaría de Hidrocarburos del Ecuador, 2015.

future oil concessions. Overall, the mapping of the territory constitutes a simplification effort in which other existing or alternative territorializations that contradict the state's assembling of Yasuní as part of translocal "resource geographies" (Bridge 2011) are downplayed or directly excluded. In other words, the petroleum map eludes complexity. According to Mol and Law,

> There is complexity if things relate, but don't add up, if events occur but not within the processes of linear time, and if phenomena share a space but cannot be mapped in terms of a single set of three-dimensional coordinates.
>
> (2002: 1)

While Yasuní is entangled in several policy frameworks that don't add up due to political, cultural, and environmental complexity, the petroleum map works to *tame* this situation by visually downplaying the various contradictions found within the national park. However, according to the management plan for the area, territorial overlaps within the national park between the ZITT, part of the Waorani territory, and several oil blocks also imply overlapping jurisdictions involving different management guidelines (see Figure 5.3). This situation creates ambiguities and a lack of coordination regarding the distribution of competences between the administration of Yasuní National Park and the oil companies (Ministerio del Ambiente 2011: 18). Besides spatial and temporal zoning, which I will discuss later in this section, overlapping jurisdictions also generate a legal void "on the ground", which, in practice, becomes a third type of zoning. Given the fact that Yasuní is distributed among a variety of policy frameworks and legal dispositions that pursue different objectives, the outcome is a legal gap that complicates certain actions while enabling others. All these *in-betweens* that constitute Yasuní have produced a series of territorial ambiguities, influencing not only the way the ITT Initiative was perceived nationally and internationally but also the way civil society organizations have articulated their efforts. Regarding this situation, an informant working with local communities in the Amazon region commented:

> I think one of the campaign's disadvantages was perhaps reducing the ITT Initiative to block 43, which made us in a way focus our attention on that space when all the other oil blocks were advancing in the rest of the area, when, for example, the major presence of isolated peoples is in [block] 16 and 14. In 2013, there was even a massacre of isolated people that occurred in an area, which was not the ITT but it was within Yasuní National Park. Of course, the first thing that produces surprise is when you are presented with an initiative to conserve a national park but then you discover that the national park has been drilled since the 1990s.

The account addresses some of the problems that arise from the lack of coherence between the various policies and territorial issues within Yasuní. In this case, overlapping zones become a distinct type of *syncretism*, which can be understood as a

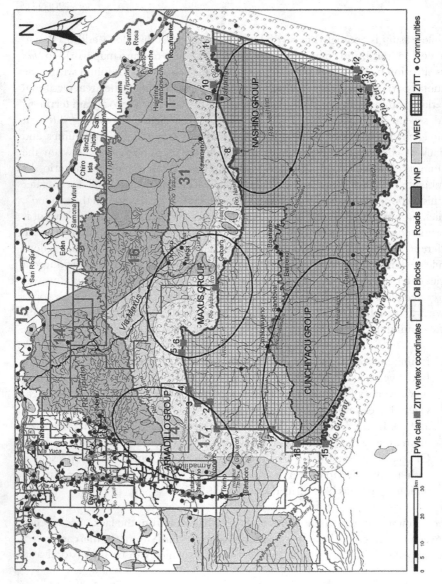

Figure 5.3 Synthesis map: Yasuní National Park, Waorani Ethnic Reserve, Intangible Zone (Zona Intangible Tagaeri Taromenane; ZITT) and oil blocks. Source: Pappalardo et al., 2013.

form of *collapse* (Law et al. 2013). In this particular mode, *purity* and fixed boundaries are not that significant. It is more a matter of being pragmatic as different practices that imply different logics are knowingly put together (Law et al. 2013). From the outside, "collapse looks confused" or "as a form of *excess*" (Law et al. 2013: 186), whereas from the inside, it can make sense as a way to handle or address complexity and a series of coexisting demands or claims from various sectors of society. While there is clearly an attempt on behalf of the state to *domesticate* the various ongoing spatial conflicts in Yasuní through *separation*, that is, by arranging the territory between different policy areas and legal frameworks, none of these cohere on the ground where they materially and symbolically meet. On the contrary, jurisdictional multiplicity and conflicting practices are manageable only "on paper", as they belong to different ministries, directorates, programs, offices, contracts, etc. Through temporal and social distribution, these actors and entities have carried out different tasks at different times (Law et al. 2013). Different governments have approached Yasuní differently, creating new zones and rearranging others.

Consequently, sectorial distribution and separation through paperwork have produced a territorial assemblage where Yasuní in practice becomes several coexisting and conflicting territories. These territories are held apart through spatial organization, but only in theory. Although the petroleum map works with spatial segregation through zoning, there are clearly overlapping territorial translations. According to the Management Plan for Yasuní National Park, the Petroleum Code of Conduct in force does not acknowledge the areas of the national park as vulnerable, only the ZITT and the buffer zone (Ministerio del Ambiente 2011: 19). This is just one example that illustrates how Yasuní National Park is performed as many different *in-betweens* in tension. In other words, what Law et al. (2013) would call a patchwork of environmental, cultural, political, and economic logics underpins the territory's spatial organization. According to the plan document, zoning is used as an instrument both to achieve proper management of the area and to identify the various actors who have a direct or indirect relationship with the park and their areas of intervention (Ministerio del Ambiente 2011: 22). The plan document lists several criteria that have been decisive for the temporal and spatial establishment of different zones within the national park:

- Physical
- Climatic
- Environmental
- Social
- Historic use

The first two types of criteria have to do with characteristics such as geology, geomorphology, soil conditions, elevations, rivers and water systems, precipitation, humidity, and temperature, etc., which are important for identifying potential risks and pressures (Ministerio del Ambiente 2011: 22–23). The last three criteria are conditions that have been more directly taken into account in the division into zones. Accordingly, the management plan emphasizes that biodiversity and ecological processes are the main factors when determining areas for sustainable resource use and corresponding management according to conservation criteria. These criteria are based on the various ecosystems and their characteristics, species, levels of threat, and endemism, interdependency between ecosystems, species, and different ecological processes (Ministerio del Ambiente 2011: 24). Strangely, ecosystems and environmental factors are, in practice, left with a *constrained field of action* despite their emphasis in the management plan, as other criteria legally incapacitate their true potential as organizing entities. More specifically, the document points towards *criteria of historic use* as a complicating factor:

> While Ecuador's Constitution from 2008 in its article 407 prohibits extractive activity of non-renewable resources in protected areas, within Yasuní National Park there is hydrocarbon exploitation due to petroleum operations that developed several decades ago and therefore have acquired rights to exploitation. Part of Ecuador's petroleum map is established on the territory of the protected area. Blocks to take into account for territorial ordering are 14, 16, 31, and ITT.
>
> (Ministerio del Ambiente 2011: 25)

Overall, *criteria of historic use* become a distinct temporal zone situated outside the management plan and its competencies. This temporal zone is constituted by a series of rights obtained over time (contracts and concessions) and materialized in a "fuzzy" petroleum map and several "oil territorialities" (Larsen 2017) on the ground. Although the Ecuadorian Constitution prohibits extraction in protected areas and zones declared as intangible, it also, and at the same time, makes this virtually possible under certain circumstances. Consequently, the article's dual agency clearly indicates that the national park status does not necessarily guarantee conservation:

> Art. 407. Activities for the extraction of nonrenewable natural resources are forbidden in protected areas and in areas declared intangible assets, including forestry production. Exceptionally, these resources can be tapped at the substantiated request of the President of the Republic and after a declaration of national interest issued by the National Assembly, which can, if it deems it advisable, convene a referendum.
>
> (Constitution of the Republic of Ecuador 2008)

This chapter will later discuss how art. 407 enabled petroleum extraction in the ITT block after the Yasuní Initiative was terminated in August 2013. Here, I want to emphasize the following: Whereas the temporal dimension tied to the oil companies' *historic use enables continued oil extraction* in the national park, it also generates a spatial zoning scheme that overlaps with *other forms of historic use* and territorial practices, which have not been given the same prerogative as territorializations evolving around oil resources. Regarding the social criteria, the management plan states:

Human ancestral presence within areas of the park, as well as processes of colonization due to the road axis within the area and in the buffer zone, are realities that have to be taken into account. The human settlements that are visible for the administration of the area are reflected in: The Intangible Zone Tagaeri–Taromenane, the Waorani Ethnic Reserve, the Waorani communities located within the protected area, and finally, the area in possession of the Kichwas on the banks of the Napo River and in the northwest of the park.

(Ministerio del Ambiente 2011: 24)

However, some of these territorializations are simultaneously deterritorializations, as they fragment ancestral ways of territorial practice or other local forms of place-making (Escobar 2001). In 1999, the ZITT was established through presidential decree as part of state policies to protect indigenous groups in the Amazon, especially the Waorani ethnic groups known as the Tagaeri and the Taromenane, who live in voluntary isolation in the rainforest. The delimitation of this zone was not finally established until 2007. With an extension of 758,051 hectares,[3] the Intangible Zone covers the southern part of Yasuní National Park and overlaps on one side with the Waorani territory and on the opposite side with oil block 31 and the ITT (see Fig 5.3). As the national park status was insufficient to keep the oil industry from expanding its operations in Yasuní, the Ministry of the Environment created the ZITT with a two-fold purpose: Prevent extractive activity from advancing in the park and, thereby, protect uncontacted indigenous groups in a separate conservation zone. However, the zoning process lacked anthropological data about the territorial practices of the Tagaeri and the Taromenane, and their place-making (Escobar 2001) efforts have frequently proved to differ from the officially demarcated area. A researcher who has worked closely with socioenvironmental challenges in Yasuní explained the creation of the Intangible Zone, emphasizing its complexity and contradictions in the following excerpt of the interview:

In 1999, shortly before Jamil Mahuad left, he signed the presidential decree by which two intangible zones are created on paper: the Intangible Zone of [Cuyabeno] Imuya with the purpose of nature conservation and the Intangible Zone Tagaeri–Taromenane aimed at protecting the isolated Indians who live there. The Zone of [Cuyabeno] Imuya had defined boundaries in the decree. Regarding the Tagaeri–Taromenane Zone, it says that within 120 days, a high-level commission will be appointed to define the boundaries. Jamil Mahuad was removed from office and the commission was never appointed, and then there was the massacre in 2003, which took place within the supposedly Intangible Zone, which was not so intangible. A group of people, in the beginning very small … we formed a citizen oversight to protect the isolated Indians. We identified as our main task achieving the Intangible Zone's delimitation and final recognition. Back then, we learned that there were plans to extend the oil blocks to what at the time was seen as the last extractive frontier, the ITT. I am referring to 2004. In 2004, we began to talk about the ITT, which was in the hands of Petroecuador. Petroecuador had plans to expand prospecting in the ITT towards the south, entering the Intangible Zone. We aimed to achieve the zone's delimitation, and in the beginning, we were not very successful because it was a rather turbulent period in the country. Every time a government fell, we had to start all over again with the negotiations, lobby, do everything at the ministerial level, the Ministry of the Environment. Finally, in 2007, we achieved the signing of the presidential decree that established the boundaries of the Intangible Zone. I wanted to tell you this because there has been strong criticism regarding the Intangible Zone. Well, from the conservationists. One of the main critiques is that the recognition of an intangible zone within the maximum category of protected areas, which is the national park that should by itself be intangible, is to waive the historical struggles of the environmental movement. National parks should be intangible. So, that is the problem. Here in Ecuador, protected areas are [protected] on paper, and on the ground, it is not respected. So, that was the big problem. If an intangible zone had not been created, by now we would have oil blocks all over Yasuní.

Me: *Do you believe that?*

I am absolutely certain. There were oil interests everywhere. If it had not been established as an intangible zone, we would not have managed to save this piece. It is a little piece of territory, but it is what remained free of oil. Why? First, because of the fear of the isolated Indians and second, because afterwards there were conflicts between sectors of the government, that is, whether to protect or not protect, if there are isolated Indians or there are no isolated Indians. In a way, it gave time to create this, which is, as I always say, like getting a foot in the door. When they are going to close the door, you put your foot there and they cannot take that away from you; it is there. Understanding this [situation], you can understand why the Intangible Zone is not the isolated Indians' territory, and it is not their territory because the criteria used to select it was not territoriality. There was never really any research done regarding where the isolated Indians moved around. From my point of view, I assume that the limits of the isolated Indians' territory are the same as the ones the Waorani had, as I told you before, the Curaray River and the Napo River.

The borders of the Intangible Zone have been shown to be rather problematic, as they do not coincide with the local and ancestral territorialities of the uncontacted groups. In other words, the ancestral territories have, on several occasions, proved to permeate and overflow the political and administrative boundaries set by the state. In fact, peoples that practice seminomadic ways of living, which is the case of the Taromenane and the Tagaeri, depend on large extensions of territory for their circuits of mobility. From an administrative point of view, this situation is challenging, as states normally operate with fixed zones and boundaries in their attempt to govern socioenvironmental conflicts. From this perspective, the Intangible Zone constitutes a political placement or grounding of the Tagaeri and the Taromenane on the map. However, it only partially manages to depict these peoples' ancestral territory, as several material signs and sightings repeatedly place them outside the demarcated zone. As artifacts that work with fixed dimensions of space, maps struggle when it comes to performing movement or more *fluid and disruptive forms of territory*. Accordingly, official maps tend to perform territories as bounded and integrated wholes and therefore fall short when it comes to enacting territorial circuits where the uncontacted groups' vital space move around with them. Hence, the Intangible Zone Tagaeri–Taromenane is a political construction that attempts to govern not only the situation of people living in voluntary isolation but also controversies revolving around the state's role regarding these peoples' vulnerability due to the presence of the oil industry, agricultural colonization, illegal logging, and environmental degradation in general. In this context, maps constitute an important political tool due to their performative capacity. What is more, in Ecuador, they have become the center of national controversies, as these "little tools" of knowledge are often accompanied by authoritative knowledge claims (Asdal 2008b: 15). The practice of mapping Yasuní and, thereby, assembling territorial realities has indeed sparked national debates regarding the state's obligations and responsibilities towards indigenous peoples living in voluntary isolation.

In August 2013, two maps that made conflicting knowledge claims were at the heart of the political debate in the National Assembly before it declared the ITT block in Yasuní of national interest, according to art. 407 in the Constitution. This process followed when the government decided to terminate the Yasuní-ITT Initiative due to low international response. To follow the trajectory and movements of this debate, we need to get close up on the two maps to see what kind of territorialities they were performing and what kind of legal linkages were enabled by these hard-working artifacts. In other words, to follow the political debate regarding the relationship between articles 57, 407, and 408 in the Constitution, we need to focus on what circulated from one site to the next, which necessarily implies keeping an eye on *form*. According to Latour, form is highly relevant as it is what allows an issue, event, or policy to acquire its mobility and connectivity. Hence, an ANT focus on form implies a shift from ideal implications to "a material description of formalism" (Latour 2005: 226). With this in mind, I zoom in on the "signs of presence" translated into maps produced by the Ministry of the Environment in 2010 and the Ministry of Justice in 2013.

The presidential request to the National Assembly concerning petroleum exploitation was accompanied by several reports prepared by the Ministries of the Environment, Non-Renewable Natural Resources, Coordination of Economic Policy, and Justice. While all the reports were distributed among the various specialized commissions in the Assembly, it was primarily the one issued by the Ministry of Justice that caused controversy among civil society actors, environmental organizations, and other stakeholders. What specifically initiated the debate was a map that rapidly emigrated from the National Assembly as it began to circulate in the media and on the internet. By relocating the areas of the uncontacted groups, this map actively worked upon territorial definitions and realities. It displayed a geographical distribution of peoples living in isolation that clearly differed from the previous map presented by the Ministry of the Environment in charge of implementing the Plan of Precautionary Measures to Protect Indigenous Peoples in Isolation [4] before the Ministry of Justice took over in 2010.

The former map, as opposed to the new one, presented four ovals (see Figure 5.4) that indicated that there had been identified signs of presence of uncontacted peoples in these areas. The map was the outcome of data collection carried out by a team of researchers that started in 2008 to map different kinds of material signs and indications of presence all over Yasuní. Different types of information were obtained through high-resolution aerial photography, interviews with the Waorani population living in the area, signs on the ground of abandoned houses, or small crops, sites where there had been registered violent attacks, etc. The team was able to identify places where uncontacted groups clearly existed at the time by combining different types of data. Based on nearby features at these locations, the groups were named Armadillo, Vía Maxus, Cunchiyacu, and Nashiño.[5] The new map created by the Ministry of Justice performed a different reality, as the oval depicting the Vía Maxus group no longer existed but had been replaced by a new oval called Tivacuno. As opposed to the Maxus group, this one was located further south outside block 16, as the oval was now placed within the Waorani territory. The Nashiño oval had been moved to an area in the south of Yasuní National Park and was no longer overlapping with block 31 or block 43 (ITT). The Cunchiyacu group was located in the same area as before, whereas the Armadillo group was no longer included in the map. Regarding this aspect, the Ministry of Justice explained that in the documents produced between 2008 and 2009 as part of the Precautionary Measures Plan, the existence of the group named Armadillo had been identified based on the hypothesis that two attacks perpetrated in 2008 and 2009 were carried out by a group of uncontacted people living in the area (Ministerio de Justicia, Derechos Humanos y Cultos 2013a). A logger and members of a *colono* family were speared and killed in these attacks. However, updated information achieved from orthophotography, the use of camera traps, and inspections on the ground indicated that there were no *malocas* (ancestral long houses) or cultivated parcels in this area. The Ministry of Justice concluded, therefore, that the attackers belonged to the Tivacuno group (Ministerio de Justicia, Derechos Humanos y Cultos 2013a). Despite this affirmation, several violent encounters over the years in the Armadillo area suggest that an uncontacted group has repeatedly been there. Material signs are significant testimonies of their presence and movements through the forest, although there are many uncertainties tied to their patterns of mobility and as to where the uncontacted peoples live for shorter or longer periods of time. In March 2013, people found a Waorani couple speared close to the community of Yarentaro located within block 16 and close to the Maxus road. This event led to a massacre of a group of Taromenane by related Waorani a few weeks later. Another Waorani couple faced a similar attack in January 2016 at the banks of the Shiripuno

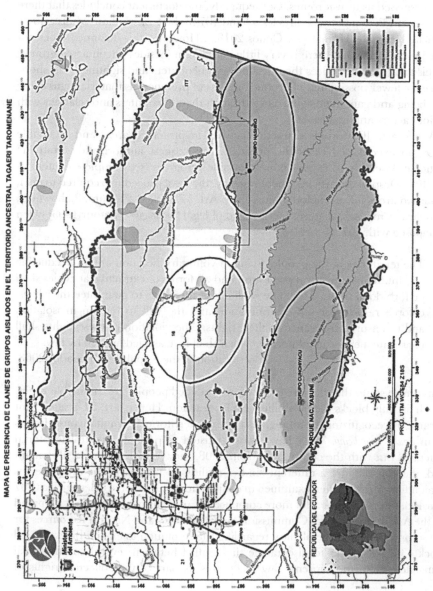

Figure 5.4 Map: Presence of isolated groups in Yasuní National Park. Source: Ministerio del Ambiente, 2010. Reproduced in Ministerio de Justicia, Derechos Humanos y Cultos, 2013b.

River, where only the woman survived (El Comercio 2016). A spear with certain characteristics is a sign that the Taromenane are in the area or have been there. However, the report from the Ministry of Justice operates with an understanding of territory as the place where malocas or crops are located; therefore, it concluded that no group is living in the area of Armadillo despite violent attacks where the Taromenane have left their spears behind. Based on information about the characteristics of the terrain, such as altitude and river locations, the report also maintains that no uncontacted peoples live within blocks 31 and 43; neither are these used as transit routes. Consequently, the document concludes that there is no contradiction between articles 57 and 407 in the Constitution (Ministerio de Justicia, Derechos Humanos y Cultos 2013a). However, this document does not take into account that there is very little population (besides uncontacted groups) in this area, which reduces the possibility of encounters and sightings. [6] In short, there are fewer possibilities of producing signs of presence, as material signs come into being and gain their significance through their encounters and relations with various actors and materialities.

While art. 407 forbids extractive activity in protected areas and intangible zones, in exceptional cases the president can request authorization from the National Assembly to exploit non-renewable resources (see complete content of art. 407). Despite the article's double agency, the legal procedure required to open protected areas for oil exploitation is clear. Art. 57 (section 21, para. 2), however, complicates matters, as it produces a kind of legal "*imbroglio*" (Latour 1993) when juxtaposed with 407:

> The territories of the peoples living in voluntary isolation are an irreducible and intangible ancestral possession and all forms of extractive activities shall be forbidden there. The State shall adopt measures to guarantee their lives, enforce respect for self-determination and the will to remain in isolation and to ensure observance of their rights. The violation of these rights shall constitute a crime of ethnocide, which shall be classified as such by law.
>
> (Constitution of the Republic of Ecuador 2008)

This article states that the territories of uncontacted peoples are intangible zones, which literally blocks the possibility of oil extraction. Hence, art. 57 is significant because by recognizing the intangibility of these territories it simultaneously brings them *legally into being*. They have always existed as part of ancestral territorial practices, but with the constitution from 2008, they move into the legal domain and, thereby, acquire a different kind of reality. At first sight, the article seems straightforward but when examined in conjunction with other articles, such as 407 and 408, things become a lot more complicated and "fuzzy". The report prepared by the Assembly's Special Commission on Biodiversity and Natural Resources for the second debate of the plenary regarding the resolution to exploit block 31 and block 43 (ITT) works to disentangle the conflict between several constitutional principles. First, the supreme law recognizes as part of the collective rights of indigenous communes, communities, peoples, and nations "to keep ownership of ancestral lands and territories and to obtain free awarding of these lands" (art. 57, section 5.). However, art. 408 declares in its first paragraph that non-renewable resources located in the underground, such as mineral and oil deposits, belong to the state and are "immune from seizure and not subject to a statute of limitations". Second, art. 407 enables the state, with previous authorization from the National Assembly, to exploit these resources even if they are located in protected areas. The report from the Special Commission on Biodiversity and Natural Resources recognizes that these principles may collide and seeks support from international law on human rights (Asamblea Nacional 2013).

Here I will not follow the legal route of the international instruments that are further enrolled in the document. My intention is to follow the trajectory of the material signs of presence (spears, long houses, trails, small crops, attacks, etc.) that were recorded, written down, photographed, and registered by the research-ers before they were translated as ovals onto maps and included in the reports that were later analyzed in the National Assembly. During the process that evolved around the declaration of national interest, the ovals on the maps bifur-cated in two different translations. While the first team working on the Plan for Precautionary Measures under the Ministry of the Environment never suggested that the four areas identified on the map should be conceived as territories,[5] the situation changed with the report from the Ministry of Justice. This situation came to have significant consequences as to what happened when the new map, which only identified three areas, encountered the legal entanglement generated by constitutional principles in tension. This refers to the conflict generated between horizontal territorialities, such as collective territorial rights of indigenous com-munities and nations, territories of peoples living in voluntary isolation, and con-servation of nature in protected areas on the one side and vertical territorialities related to non-renewable resources located in the underground on the other side.

The Special Commission on Biodiversity and Natural Resources acknowledged the possibility of exploiting petroleum in protected areas in exceptional cases and at the same time the need to protect peoples in voluntary isolation; therefore, it recommended not exploiting natural resources in the zone declared as intangible in favor of the Tagaeri and the Taromenane (Asamblea Nacional 2013). In short, to leave oil extraction out of the ZITT. In practice, the recommendation *enacted the Intangible Zone as the territory of the uncontacted peoples*, and the map from the Ministry of Justice became an important *ally* as it displayed the areas of presence outside the boundaries of the oil blocks in question. While the Minister of Justice emphasized that the map presented to the Assembly was based on updated information, sev-eral civil society actors questioned its legitimacy. This situation can be observed in the following statement provided by a former member of the Constituent Assembly, who emphasized that Yasuní-ITT enjoyed a double protection due to environmental considerations and because of the uncontacted peoples. According to the informant, the National Assembly can legally overcome the intangibility of natural protected areas, putting art. 407 to work, but it cannot legally overcome the intangibility of cultural areas. In other words, art. 407 has no authority over the content of art. 57:

> The Assembly's authorization is unconstitutional, not because of the environmental issue, but because peoples in voluntary isolation live there. The Assembly did it anyway. The political and democratic system is designed to solve the issue, because we are not governed by angels. It is not that serious that the Assembly did this. It is serious that the President made the proposal and it is even more serious that the Ministry of Justice that was in charge of the plan for protecting the peoples in isolation changed the map so the Assembly had the justification to do it.

The comment emphasizes the active role of the map in the legal controversy related to extending oil extraction to new blocks in Yasuní. According to the informant, the map supported a particular understanding of territory as a fixed or bounded place where people live and, thereby, provided the members of the National Assembly with a legal justification to open the new areas to petroleum activity. However, this understanding differs from how, for example, the International Labour Organization (ILO) Convention (1989) defines territory, where the concept refers to "the total environment of the areas which the peoples concerned occupy or otherwise use" (Art. 13, 2.). How the different ways of translating "areas of presence" intersected the political debate evolving around the ITT block and turned out to be highly relevant for the outcome can be observed in the following statements provided by a member of the National Assembly from Alianza País[7]:

> The information we requested was how much will the park's natural area be affected and, most importantly, that none of these interventions would undermine the rights of the inhabitants of the area. Nevertheless, it does not stop being controversial and additionally, when you talk about people, about human beings, it is very difficult to say they are located in this circle; they will not go beyond this circle. Even more so, when we know the logic of life of these peoples, peoples that circulate in these areas. So, several conditions were established. One was highly publicized: About intervening less than 0.1% of the park taking into consideration the most adequate techniques of oil extraction. The whole thing about transportation using helicopters, ecological trails. So, pretty assured about that. There was really a lot of information and the maps of the peoples, and frankly I have to say that from the state's position the maps were given to us saying that there were no settlements. There had been no sightings of settlements of these peoples and nationalities, which made the situation clear to us from the state's perspective. Nevertheless, organizations or collectives articulated in Yasunidos said the opposite.
>
> Me: *So, there are no confirmed sightings in these blocks until now?*
>
> "Sighting" is a very relative word. With a settlement you have a series of houses, cabins, tools, certain things, but a sighting can be a spear or it could be a trail marked by three or four [people] passing. So, a sighting can be used

in any way to say "Oh, it is only a spear" or you could say, "It is actually a spear". That is how politics are handled in a performative manner. But we said that in case of a sighting, the activities have to be suspended until they comply with all the protocols.

This informant emphasizes that the state's perspective was always clear to the members of the National Assembly. In other words, it was clear that the map from the Ministry of Justice enacted the ovals as areas where indications of settlement existed. The example with the spears is interesting, because it illustrates how "things" are performed and become different realities. The state's focus on more solid objects (e.g., houses) to establish territorialities of uncontacted groups is rather contradictory, when we know that these peoples' lives and subsistence depend on trekking in the forest. Their cyclic movements also imply returning to the place of their ancestors every three or four generations (Colleoni and Proaño 2010), which responds to a logic of regenerating the ecosystems on which they are highly dependent (Álvarez 2016). Although the Ministry did not find any signs of presence in blocks 31 and 43 during the period of data collection, uncontacted groups could potentially enter the area in the future. This situation shows some of the contradictions and uncertainties that arise from the multiple logics that operate in the assembling of the territory. All spatial zoning within Yasuní, even the borders of the national park itself, are exogenous to the region and local territorialities, as they have been designed based on political and economic requirements largely found outside of Yasuní. In other words, many of the state's territorializations are performed according to translocal dynamics. The current overlapping zones within Yasuní National Park thus reflect the "spatial dimensions of power and politics" (Müller 2015: 27).

Besides the creation of the ZITT, the history of the Waorani territory is another illustrative example that responds to a heterogeneous assembling process based on spatial requirements of transnational oil companies, national economic policies, and an ancestral spatiotemporal ordering of nature and humans. Before a group of missionaries from the Summer Institute of Linguistics (SIL) contacted the Waorani in 1956, they lived as seminomadic hunters in a territory that extended from the Napo River in the north to the Curaray River in the south, a territory of approximately 2 million hectares of tropical forest (Franco 2013: 145). Their history of contact is closely related to the exploration and exploitation of natural resources in Amazonia, especially petroleum. During the 1940s, the Shell Oil Company conducted surveys deep in the Ecuadorian Amazon (Martz 1987: 52). Although the company drilled several exploration wells, it never began production of these reserves, arguing that "the Oriente lacked commercial petroleum" (Martz 1987: 52). Shell decided to return a large part of the concession to the Ecuadorian government in 1948, due to supposedly low profitability. However, the presence of the Waorani, who fiercely defended their territory and resisted all forms of contact, was an equally decisive factor that made the oil company abandon the project (Franco 2013: 146).

After establishing contact, the missionaries from SIL initiated a process of relocation, as they moved the majority of the Waorani (though some clans resisted) to a protectorate they founded at the headwaters of the Curaray River in Tiweno, leaving the area "open" for the international oil companies and agricultural colonization. As discussed in the previous chapter, both processes were given high priority by the Ecuadorian authorities. The relocation of the Waorani resulted in increased sedentarization as compared to their traditional economy based on hunting and collecting, which necessarily meant moving across large extensions of land and changing residential location from time to time. Rival (2015: 65) emphasizes that these cycles of movement and activities in the forest translate into small interventions through which the Waorani enrich their habitat as they contribute to the accumulation of species in specific areas, for example, by leaving behind a considerable amount of seeds before moving to another location. Hence, the forest exists at present due to the activities and work carried out by other humans in the past: By interacting with the forest, a social relation that extends across generations is being produced (Rival 2015: 66). The spatial and temporal aspects of territorial practice based on trekking in the forest shows that territory is "simultaneously biophysical and epistemic space where life is enacted according to a particular ontology"[8] (Escobar 2015: 35), where humans and non-humans collaborate and establish a variety of productive relationships.

The SIL missionaries played a significant role in the process of "pacification" of the Waorani and enjoyed considerable goodwill among successive Ecuadorian governments. This situation changed with the Roldos administration, which became increasingly critical of SIL's activities and decided to expel the organization from the country in 1981, which in practice ended the existence of the Waorani protectorate (Franco 2013: 154). When many Waorani families returned to their place of origin, the territory had radically changed due to new territorializations produced by the state, which were primarily based on the requirements of an expanding oil industry. Both the creation of Yasuní National Park and the Waorani Ethnic Reserve were direct consequences of increased petroleum activity in the area (Rival 2015: 26). This situation became evident in 1990 when the government of Rodrigo Borja officially granted the Waorani 612,650 hectares in addition to the 66,570 hectares previously awarded in 1983 (Rivas and Lara 2001: 35–36). The official document that recognized the new extension of the Waorani territory (which is only a part of their original territory of approximately 2 million hectares) stated that in accordance with the Constitution, the Waorani were only owners of the surface, as the state was in control of the underground (Rivas and Lara 2001: 37). The document therefore specified that the Waorani could not oppose or obstruct exploration and extraction of minerals or petroleum resources in their territory (Instituto Ecuatoriano de Reforma Agraria y Colonización (IERAC) 1990 in Narváez 1996). This is another example of how vertical territories impact the ordering of space above the ground: Boundaries are constantly being negotiated as pipelines and the corresponding road axes have reconfigured the previous frontiers and territories.

As part of the territorial syncretism that characterizes Yasuní, we also find several coexisting and overlapping natures. On the one hand, there is what Escobar calls *capitalist nature*, which refers to a commodified nature that is taken into account primarily as natural resources; on the other hand, we have what he refers to as *organic nature*, which is entangled in culture and local knowledge and where the modern division between nature and culture tend to be non-existent (Escobar 1999: 6–9). In Yasuní, capitalist and organic natures constantly collide and at the same time, they reconfigure each other. Accordingly, the struggle over Yasuní has to do with "politics of nature" (Latour 2004), as there is no single, coherent, or all-encompassing nature involved, but several overlapping and partially connected natures. On the other hand, it is also a matter of "politics of place" (Escobar 2001) in which local territorialities are entangled in the national and international ordering of space. Against this background, the ITT Initiative launched by the Ecuadorian government in 2007 is significant, as it constituted a political strategy that could potentially subvert the territorial dynamics in Yasuní in terms of establishing a new order between the underground and the surface. With this policy shift, the petroleum deposits located beneath the ITT block were translated as "carbon" (a new commodity), which is subjugated to the cultural and biological diversity of horizontal territories. The next section examines the first moment of translation as the proposal enters the government's political agenda.

Transition from civil society to government

The proposal to leave the oil underground in the ITT field entered the government arena in early 2007 when Alberto Acosta, an economist and academic, was appointed Minister of Energy and Mines in President Correa's first administration. The new minister, who was known as a critic of the traditional development model based on petroleum extraction, had close ties to national environmental organizations. Together with key actors such as Esperanza Martínez from Acción Ecológica, Acosta had participated in discussions related to the proposal of an oil moratorium and the required transition towards a post-petroleum development in Ecuador. Originally, the idea was to stop the oil frontier from expanding further, as the proposal included not only Yasuní-ITT but also the whole central-south of the Ecuadorian Amazon. Overall, the idea was to halt the opening of new oilfields. In the past, this had been politically difficult to achieve due to limited proven petroleum reserves and a declining production since 1993 in Ecuador's most important oilfields (Larrea and Warnars 2009: 220). As Minister of Energy and Mines, Acosta became an important link between aspirations and ideas that had originated within civil society and the government. Hence, Acosta's appointment as head of the Ministry in charge of petroleum policies constituted a decisive step that made the environmental issue part of the government's political agenda, despite the economic opportunity tied to the ITT field due to the high oil prices at the time (Álvarez 2013: 83). Nevertheless, the proposal to strand petroleum assets in Yasuní in exchange for contributions from the international community encountered several obstacles and setbacks on its way to institutionalization at the national level. During the initial *problematization* (Callon 1986), tensions surfaced between the administration of Petroecuador interested in promoting a partnership agreement between Sinopec (China), ENAP (Chile), and Petrobras (Brazil) to exploit the ITT field and the Minister of Energy and Mines, who favored keeping the oil underground and finding a mechanism for international cooperation (Martínez 2009: 14). Despite political difficulties and a problem definition that pointed towards oil extraction and, thereby, the petroleum industry as an essential part of the environmental issue, the ITT proposal managed to gain adherence and enroll important actors within the government, including President Correa. The board of directors of Petroecuador officially announced the resolution to keep the oil in the ground at the end of March 2007. Shortly after, the Ministry of Energy and Mines sent out a press release declaring that the first option, or plan A, was to forego the petroleum deposits in the ITT field, which is Ecuador's "second largest untapped oil fields" (Bass et al. 2010: 2) and estimated to hold 846 million barrels of oil (Larrea et al. 2009). This decision depended on the international community's participation in compensating Ecuador with at least half of the revenues that the oil resources would generate if commercialized on the international market. If the goal was not achieved, the country would move on to plan B, which meant drilling for oil to obtain the resources the country needed to finance development. Leaving the second option open in case the international community decided not to support the Yasuní-ITT Initiative was regarded as a strategic mistake, which generated considerable criticism, especially from environmental organizations, as some stakeholders believed the mechanism presented by the government sounded more like blackmailing than asking for contributions. Within civil society, many feared that the intention to exploit the ITT block in the absence of financial compensation could overshadow the initiative itself, as it transmitted mixed signals to the international community. In rather ironic fashion, a former member of the Administrative Committee for the Yasuní-ITT Initiative commented:

> When you have a plan A and a plan B it is [as if] I go to my girlfriend and I tell her do you know what, I want to marry you. You are my plan A but I have another girlfriend who is my plan B. Listen, would it not be a good idea to eliminate plan B first and decide if you want to marry me before you propose. Any proposal I make has to be clear and exclusive, not halfway. All the signals the President gave were completely contradictory.

Things were perceived differently inside the government: A plan A also required a plan B in case the international community decided not to support this novel mechanism, which had never been implemented before and that directly linked non-extraction of fossil fuels to the preservation of biodiversity, indigenous rights, and climate change mitigation. To forego oil revenues is by no means an easy political decision, especially for a country where oil revenues constitute one of the state's most important income sources: Oil production contributes to around 35% of the national budget (Martin 2015: 120). From this perspective, "it is surprising that the proposal existed in the first place, and striking that it was implemented, even if only for a small window of time" (Sovacool and Scarpaci 2016: 169). A former minister of President Correa's first administration stated about this situation:

There are many people who complain, and they have complained to me say-ing why did we have option A and option B. Option A is to leave the oil in the ground and option B is to extract the crude. If there had not been an option B there would not have been the option A. The fact that there was the option A to keep the oil in the ground was already a victory, because the normal thing to do in this country was to extract all the oil possible. Of course, it is difficult not to exploit these resources, and then you have the discourse of the President and others that we are like beggars sitting on a sack of gold because it is money, it is the future.

The comment indicates how challenging it was for the government to land on option A. However, after several years of campaigning and attempts to enroll contributors and gain international recognition and support, the government decided it was time to give up in August 2013. Six years had passed since President Correa launched the proposal of leaving the oil in the soil during the High-Level Dialogue on Climate Change in the General Assembly of the United Nations in September 2007. Correa's speech emphasized that the Yasuní-ITT Initiative was a significant contribution from a developing country to global collective action, which demanded co-responsibility as a means of addressing environmental degradation and global warming.[9] When the government decided to withdraw the proposal and finally liquidate the trust fund due to the lack of contributions, the alternative strategy or plan B was set in motion. The policy shift resulted in huge disappointment and opposition among different civil society groups that considered the ITT Initiative a long-needed opportunity to leave behind the reliance on petroleum exploitation and pursue a more sustainable model of development. In other words, this had been Ecuador's *Kodak moment*.

Correa's position regarding the extraction of natural resources was pragmatic, stressing that "we cannot be like beggars sitting on a sack of gold". The main argument was that the country needed economic resources to improve living con-ditions for those marginalized groups that have been excluded from economic growth and who live below the poverty line within Ecuadorian society. The for-mer government's conception of well-being and development has been widely debated within various sectors of society. This conception also became a cen-tral element in the process around the Yasunidos movement and their campaign aimed at achieving a national referendum regarding the future of Yasuní. The chapter's final section examines this point.

Carbon translations

When analyzing fieldwork data, I spent considerable time trying to trace the vari-ous ideas, proposals, policies, processes of resistance, etc.; in short, all the various initiatives that were woven into what finally came to be known as the Yasuní-ITT Initiative. My intention was to look into moments of translation that moved the initiative from one phase to another, which in practice meant looking at how the proposal traveled between several construction sites, expanding the network as new actors and actants were *enrolled* and *mobilized* (Callon 1986). I eventually realized that there were several competing network translations going on simultaneously that probably affected the initiative's attempt to go global. As I see it, the heterogeneity of the ITT Initiative actually makes it more appropriate to talk about several initiatives that were congruent on some occasions but became competing rival translations on others. The tension between conflicting ideas can be understood as an effect of parallel dynamics working within the ITT Initiative. Overall, it constituted an incipient step towards the *ecologization* of collective life between humans and nonhumans (Latour 2007). By turning the spatial and territorial dynamics in Yasuní upside down, the initiative pursued a more harmonious and friendly relationship with nature. According to Latour, ecologization is a process that implies "recognizing the ever-more complex interweaving of humans, animals, ecosystems and technologies, which is made evident by our current ecological crises" (Blok and Jensen 2011: 169). From this perspective, the 2008 Constitution and, particularly the Rights of Nature (art. 71–74) came to play a significant role as a legal fundament and, maybe more importantly, as an *ethical guarantor* for the initiative. Another important constitutional principle, which I briefly referred to in the previous chapter, is *Sumak Kawsay* (in Spanish *Buen Vivir*) that in Kichwa means living well, or life at its fullest. The concept is used to refer to a harmonious way of living in community and interacting with nature. According to Lalander, writing on neo-constitutionalism, "the indigenous ethical and philosophical conception of *Buen Vivir/Sumak Kawsay* is the backbone of the new Ecuadorian Constitution" (2014: 627) and, therefore, constitutes an important element in a larger political debate on what comprises true development. It can be fruitful to see the intertwined proposals and aspirations that were translated into the Yasuní-ITT Initiative in the light of this changing political landscape, where the relationship between society and nature was taken into account in an unprecedented way. As such, the inscription of the concept of *Sumak Kawsay* and the *Rights of Nature* can be seen as an expression of what Escobar (2011) calls "transition discourses", resulting from the global crisis that is largely provoked by modern dualistic practices in which economic logics have separated and reduced nature to an instrumental role within production. From this perspective, the Yasuní-ITT Initiative constitutes an important example of how these discourses occasionally materialize and are acted upon in practice. Nevertheless, the Yasuní-ITT Initiative was fundamentally a hybrid project, as it also drew on *modernization*: Nature was "distillated" through a series of numeric translations that enacted Yasuní as a unique, pristine, and untouched place. On the other side, Yasuní was translated as carbon: The world would avoid 407 million metric tons of CO_2 emissions by not extracting the fossil-fuel reserves located underneath the ITT block. Another 800 million metric tons of CO_2 would remain unreleased due to avoided deforestation.[10] Interestingly, these translations show how Yasuní-ITT moved from one carbon economy to another. The ITT block comprised an important part of Ecuador's proven petroleum reserves. These reserves were translated into avoided CO_2 emissions to gain adherence within the international climate regime and stabilize the network. In other words, Yasuní moved from what Bridge (2011) has called the "old" to the "new" carbon economy. This transition was a step away from how civil society organizations had articulated their proposals in the past based on conceptions of ecological justice and the responsibility for historic emissions placed on developed countries. Overall, Ecuadorian environmental organizations wanted to operate outside the carbon market, as they believed it was just another way of commodifying nature and an attempt on behalf of the industrialized countries largely responsible for the climate issue to commodify the atmosphere. The following reflection provided by a representative from Acción Ecológica problematizes the way carbon was used as the unit of translation in the Yasuní-ITT Initiative:

> In a way, the problem from a symbolic viewpoint originated when the idea of the equivalence between oil and carbon was established. I think that was a big mistake that we did not realize in due time, because the oil and the exploitation of Yasuní was so much more than protected carbon. It was the water, the biodiversity, the peoples. Of course, the moment the initiative was founded on carbon as the exchange unit, the conditions were created so that it could enter the market, which was exactly what we did not want. If we had continued just talking about oil, it would have been better. In addition, carbon is a virtual thing. I do not understand why people think carbon is more comprehensible than oil, as nobody sees carbon. Oil is something that you see, you burn it, you put gallons in your car, well, you see it. But all of a sudden everything is transformed into carbon. So, it seems like what Yasuní does, is that it saves 404 million tons of carbon and that is not it. It was going to prevent oil spilling, rivers from being contaminated, and the destruction of biodiversity.

The informant emphasizes that the Yasuní Initiative originally provided a way to question the efficiency and ethics of carbon markets. In the interview excerpt, carbon is portrayed as a form of reductionism: As the unit of translation, carbon overlooks many of the complex interconnected issues that are tied to oil exploitation. While the initiative had a holistic approach, the informant believes that the transformation of the oil reserves into avoided CO_2 emissions, in other words framing Yasuní primarily as a carbon-emissions problem, drew attention away from all the other matters of concern that were involved. While the Yasuní Guarantee Certificates (Certificados de Garantía Yasuní; CGYs) were not officially carbon credits but legal obligations that indicated the amount of avoided CO_2 emissions, they could in theory acquire a different kind of agency and, thereby, extend their applicability by becoming carbon offsets. In *Yasuní-ITT Initiative: A big idea from a small country* (Larrea et al. 2009), a policy document that explains the scientific and economic rationale behind the proposal, the *plasticity* that is worked into the CGYs can be observed when comparing the following paragraphs:

> In exchange for the contributions, the Ecuadorian State will guarantee to maintain ITT oil reserves underground indefinitely. The government will issue guarantee certificates for the nominal value of the compensations

(Yasuní Guarantee Certificate – CGY), up to the quantity of 407 million tons of carbon dioxide not emitted. The real backing for the guarantees will be the value of the investments made by the capital fund.

(2009: 4)

Ecuador proposes to countries that are sympathetic to the Yasuní-ITT Initiative to participate with contributions to the fund, including, if applicable, debt for conservation swaps. A second option, only in the case of North America if a cap-and-trade system is established, is the formal recognition of CGYs as carbon credits, and their integration as a pilot scheme, under specific conditions: These certificates could be 1) purchased directly by governments, or 2) purchased by companies, but subject to the condition that the CGYs will be considered within the total quota of annual emissions permits.

(2009: 4)

Some civil society organizations objected to these "carbon translations", as they could potentially link the ITT Initiative to a North American carbon-offsetting scheme, if it was created, which would insert the CGYs into the market. Although this situation was rather hypothetical, it caused tension and controversies between civil society actors and the government as some stakeholders stated: "The initiative had become [the] mercantilization of nature and commercialization with carbon credits", which made the involvement of civil society very complicated since their involvement required *a purist approach*. On the other side, the Administrative Committee for the Yasuní Initiative, appointed by the government to develop the proposal, believed that the first problem definition presented a series of shortcomings that left the initiative with reduced international viability. While the government was initially enrolled by civil society and then decided to move forward with the proposal, it also decided to reframe the issue. Consequently, the Administrative Committee applied a different focus in order to achieve *interessement* (Callon 1986) among international actors in accordance with a slightly different problem definition:

When I received the proposal, it was not a strategic proposal, it was a "romantic proposal", a proposal about socioeconomic and global justice, North-South and the whole idea. The Europeans had contaminated since the Industrial Revolution in the mid XIX century until 2000. They had used the atmosphere as a garbage bin, so they had to pay. That was the theory behind [the initiative]. One way to pay is that we have the right to contaminate just as much as they did to get even in contamination. The concept is a bit silly but that was the idea related to justice, equity. Since we do not intend to contaminate, they should compensate us for this effort. I am simplifying, almost caricaturing, but that was the idea. My idea, and here there is a strategic logic: Climate change "is on its way" no matter what the Republicans say. They are fools, and climate change is coming. Fifteen years from now, oil will be like a poison. It will be considered waste and it will be necessary to leave it in the ground. So, we will get to the point where we cannot burn more oil because we will reach the 2°C limit in temperature rise. Eventually, we will have to stop. Our proposal was that we should begin with those countries that combine two elements: They have oil and above the area where they have oil, there is exceptional biological value. There are many places with these characteristics around the world where you have national parks, where there is a rich biodiversity and underneath there is oil. When you start analyzing there are more than 40 places that have these characteristics. Let us give those countries the option of using the Yasuní model. That was the idea.

Less focus on ecological justice in favor of an increased emphasis on the necessity of keeping the oil in the ground to accomplish the political goal of not surpassing a 2°C rise in temperature is a translation that implies a movement closer to the accounting logic that underpins the Kyoto Protocol. However, there is a substantial difference: Whereas the Kyoto Protocol operated with the concept of reduced emissions using the year 1990 as a baseline,[11] the Yasuní-ITT Initiative introduced the concept of *net avoided emissions*, thereby shifting the focus from reduction of greenhouse gas (GHG) emissions to avoidance through non-extraction of fossil fuels. Overall, the transition from civil society to government produced a displacement, which necessarily constitutes a form of *betrayal*, as network translations are never just faithful representations (Law 2003). On the contrary, when "transfer" takes place there is also change, shift, and movement (Law 2003). Although the process that evolved around the Yasuní Initiative is full of overlapping *origin stories*, there are also *accounts of emigration* that I intend to trace in the next section.

(Re)assembling the Yasuní-ITT Initiative

When the ITT proposal finally became an official government policy, the next challenge was to develop a mechanism that better responded to the complexities of the Ecuadorian Amazon than the existing international frameworks and instruments. This was not an easy task, as many international actors found it to be counterintuitive for a developing country to leave hydrocarbon resources untapped. The Yasuní-ITT represented approximately 20% of the country's total oil reserves (Larrea et al. 2009: 10). Several informants who worked directly with the design of the Yasuní model have emphasized how challenging it initially was to gain acceptance for a conservation initiative that tried to solve several inter-related and entangled issues that did not *fit smoothly* within already existing climate mitigation policies. The Kyoto Protocol worked with accounting practices concerned with reducing emissions through International Emissions Trading, Joint Implementation, and the Clean Development Mechanism (CDM), and the Yasuní-ITT Initiative was not directly compatible with any of these options.

The Kyoto Protocol assigned the Annex B countries specific amounts of GHG emissions allowed during the established commitment periods. Any additional emissions could be purchased from countries that had emission units to spare.[12]

In other words, countries that were not going to use their whole quota could sell the remaining units to other countries. In order to make the system work a new commodity was produced, namely GHG emissions, with a corresponding market. Since carbon dioxide is the main GHG, it became the unit of translation used to make all greenhouse gasses commensurable (as CO_2 equivalents) despite having different chemical compositions and global warming potential. Similarly, Joint Implementation was applied exclusively to Annex B parties, which left Ecuador with the CDM as the only mitigation strategy that directly included developing countries' participation. The CDM, however, does not consider avoided emissions from not extracting fossil fuels. When emissions are avoided in one place, this evidently contributes to reduced emissions globally. Nevertheless, in the UNFCCC scheme, Yasuní would become an accounting problem. How specifically would these avoided emissions be accounted for and where? In short, the main problem was how to take Yasuní into account when it was neither directly anchored in any of the established mitigation mechanisms of the Kyoto Protocol nor did it constitute a Reducing Emissions from Deforestation and Forest Degradation (REDD) initiative. An informant who participated in designing the Yasuní mechanism explained some of the difficulties the initiative encountered on its journey from national to international arenas:

> REDD simply talks about reduced emissions from deforestation and degradation, but the oil issue is not treated at all. So, having included the idea of leaving an oil reserve without exploiting it indefinitely is entirely new. As with all new ideas, and in this case, it was radically new back in 2007, now many people talk about it. Since it was entirely new, we could not find any mechanism in which it could fit. It could have been an obstacle, but in fact it wasn't. Let's say it was an obstacle possible to overcome. Our original idea and the idea a group of people had was we can issue a kind of certificate that later can enter some kind of mitigation mechanism, for example, the Kyoto Protocol. As you know, the Kyoto Protocol works based on certificates of reduced emissions on the European carbon market, and for developing countries it has a mechanism called Clean Development Mechanism. When going through the legislation of the CDM there was not any possibility of turning Yasuní into a CDM case because first, CDM does not even include REDD, just reduced and not avoided emissions. So, you have reduced emissions when an existing plant that contaminates, let's say, a thermoelectric plant, is replaced. I close the thermoelectric plant and replace it with a hydroelectric plant, for example, or solar or wind … So, there are reduced emissions and for those reduced emissions there is a mechanism through the carbon market within the Kyoto Protocol, which allows this country to receive a contribution. That was not possible because we were not reducing emissions but avoiding them, going to the root.

An entirely new framing process was required, as neither REDD nor the Kyoto Protocol was compatible with the Yasuní-ITT Initiative and clearly presented limitations for a *problematization* that strived to go underground to the origin of the problem. REDD only addresses deforestation and forest degradation as surface solutions while the Kyoto Protocol focused on reducing emissions at the end of the production and consumption chain. None of these frameworks could address the environmental, cultural, and political challenges of the Ecuadorian Amazon, as these issues are clearly entangled and rooted in the subterranean. By keeping the oil underground, the Yasuní-ITT Initiative aimed to contribute to climate change mitigation, protect peoples living in voluntary isolation, and conserve the exceptional biodiversity of the rainforest. Accordingly, contributions to the Yasuní-ITT Trust Fund, set up in 2010 together with the United Nations Development Programme (UNDP), would "enable the Government to face the challenges of climate change and sustainable development by changing the country's energy matrix, through investment in renewable energy, environmentally-friendly projects such as hydroelectric, geothermal, solar, wind, biomass and tidal plants" (Yasuní-ITT Trust Fund 2011: 1). While this Capital Fund Window was oriented towards financing renewable energy, in order to change the country's energy mix and enable a transition away from an extractive development model, the Revenue Fund Window was oriented towards development programs, reforestation, conservation of ecosystems, research, and technological innovation.[10]

The ITT Initiative, as a new approach towards climate change mitigation with the ambition of being replicated in other megadiverse countries with significant fossil-fuel reserves (Larrea et al. 2009: 5), necessarily had to forge *alliances* that allowed it to move forward, circulate among potential participants, and enroll them into the network. Accordingly, policy documents became important sites where Yasuní was assembled politically, but also scientifically through data about the greenhouse effect, global warming, climate change, geological periods, and numerous taxonomies that certified and consolidated Yasuní as a biodiversity "hotspot". To construct the ITT Initiative as a common global good, scientific knowledge consistently worked to "break down Yasuní in numbers". In other words, Yasuní was made accountable. The following paragraph from *Yasuní-ITT Initiative: A big idea from a small country* (Larrea et al. 2009) is an illustrative example:

> Scientists agree on the park's unique value due to its extraordinary biodiversity, state of conservation and cultural heritage. The reserve has an estimated 2274 tree and bush species, and 655 species have been counted in just on hectare: more than the total number of native tree species in the United States and Canada combined. The park has 593 recorded bird species, making it one of the world's most diverse avian sites. There are also 80 bat, 150 amphibian and 121 reptile species as well as 4000 vascular plant species per million hectares. The number of insects is estimated to be 100,000 species per hectare, the highest concentration on the planet. The species found in the park have a high level of endemism.

> (2009: 13; emphasis removed)

As a strategy to produce valuable nature and politics for preserving this nature, the national park was drawn together as an entity through a series of numeric translations that circulated in the emerging network. In this context, I echo Asdal, who emphasizes the importance of tracing how "the enactments of nature and the enactments of economy go together" (2008a: 124). In the case of the ITT Initiative, Yasuní is taken into account both as an environmental question, where valuable nature is being produced through quantities and categories, and as an economic dilemma tied to development and the need to transit towards "a post-fossil-fuel model".[10] As part of making Yasuní politically robust, policy documents relied heavily on scientific knowledge and a series of facts about carbon sinks, petroleum reserves, endemic species, global warming, etc. As the initiative moved through a series of new network translations, it increasingly required *strong scientific allies* that supported the thesis of keeping the oil in the ground. The importance of scientific knowledge systems can be observed in the following account provided by one of the previous members of the Administrative Committee for the Yasuní-ITT Initiative:

> Back then the evidence did not exist but later the thesis of the 2°C limit of climate change, which could prevent a planetary catastrophe, was accepted. Maybe it is not a certain limit but it is a reasonably certain limit of global warming and this limit was associated with a concentration of carbon dioxide of 350 ppm, which is, for example, the Stockholm Environment Institute's planetary boundary. The idea was, well, how does this relate to future emissions? And Meinshausen's work, which was published in 2009 in the journal *Nature*, indicated that we have to leave at least half of the currently proven fossil-fuel reserves to meet this goal … and in 2012 they published the same thing. They expressed with absolute clarity the same thesis, but [now] it was the International Energy Agency (IEA), an international European organization. It was not a person or a scientist but an institution. So that consolidated the thesis a lot. At present, there have been other pronouncements, including a reference I think in the last Intergovernmental Panel on Climate Change (IPCC) report. So now, we are becoming mainstream, which means that before a little and "absurd" country like Ecuador showed up saying this, now there is almost consensus, at least scientific consensus. Now it is a lot more conventional and people talk more about the idea of keeping fossil fuels in the ground.
>
> I think the conception of the issue changed radically because Yasuní [Naional] Park was recognized in a very serious scientific article that was published in 2010 in *PLOS ONE*, which is maybe the other huge event that changed the situation. This article was written by a considerable number of biologists who described it as the place with the greatest biodiversity in the western hemisphere. So, that was another important scientific recognition of which we did not have evidence, at least not written. We knew that we were in that category, but we did not have a high-quality product recognized worldwide, a publication that confirmed it.

According to Latour, when controversies intensify, they also tend to become technical. Disagreements can end up opening a series of black boxes that lead the controversy "further and further upstream, so to speak, into the conditions that produced the statements" (Latour 1987: 30). In the case of Yasuní, we are literally taken *upstream* as the initiative opens up the black box of global climate policies that frame the climate issue as an emissions problem, whereas the Yasuní-ITT Initiative pointed towards upstream oil operations as the site where the climate issue begins. The Yasuní thesis constituted a comprehensive idea that required a movement back to the origin of the climate issue and focusing on the extraction of fossil fuels as the real *matter of concern* (Latour 2005). When debates become heated, local resources and statements "are not enough to open or close a black box. It is necessary to fetch further resources coming from other places and times" (Latour 1987: 30). Documents, research, and acknowledged scientific statements that have already managed to become solid facts are important allies as they bring with them authority that can help settle a controversy. When several actors and organizations began linking the possibility of reaching the 2°C objective with the necessity of leaving a major part of known fossil-fuel reserves in the ground, Yasuní, in theory, became a stronger proposal. Consequently, statements from a scientific article in *Nature* or organizations such as the IPCC and the IEA added new allies to the proposal of leaving the oil in the soil. In this context, the article "Global Conservation Significance of Ecuador's Yasuní National Park" (Bass et al. 2010), published in *PLOS ONE*, was invaluable, as it not only came with a considerable group of authors, but it also contained references to other scientific publications, which were simultaneously enrolled as they added support for Yasuní's uniqueness. Consequently, the article participated in how Yasuní *came into being as a megadiverse site.*[13] However, one controversy can always open new ones. Norway was one of the countries that the Ecuadorian government unsuccessfully tried to enroll in the network. The Norwegian authorities, however, declined to participate, as the proposal did not explicitly commit to reducing emissions in Ecuador. The possibility also existed of intensifying extraction outside Yasuní, ending up with the same carbon volumes due to carbon leakage. Another critical point was setting a precedent by compensating a country for leaving oil underground, as other countries, like Saudi Arabia, could demand the same thing. However, according to the Yasuní-ITT documents, the idea was to replicate the Yasuní model only in developing countries with rich biological diversity. According to a representative from the Norwegian Ministry of Climate and the Environment, biodiversity is not necessarily a uniform dimension:

> Biodiversity is not an unambiguous dimension. Most countries believe they have, if not such [a] high richness, something they believe is very valuable and unique for their country and valuation of biodiversity is almost impossible. So, the discussion among countries about who has the most important biodiversity is almost impossible. At least if it comes down to political acceptance, it is possible to think that we could have some objective standards, but there is still a long way to get there.

The comment illustrates that nature and nature issues are coproduced by science and politics. Consequently, nature is by no means an "apolitical" entity (De la

Cadena 2010) but an important means of negotiation. What was at stake in this case was an entirely new approach towards climate change mitigation and development, linked together by rich but vulnerable nature. The Yasuní-ITT Initiative attempted to shift the main focus of climate change mitigation from the atmosphere to underground by broadening the scope from reduced emissions to include avoided emissions obtained by preventing carbon reserves from being released.

Overflowing the Kyoto Protocol

While framings constitute attempts to settle controversies, frames are never completely stable. There is always the possibility of interaction and exchange between what is inside the frame and what operates outside its boundaries. Accordingly, "any frame is necessarily subject to overflowing" (Callon 1998b: 18), as the very process of framing contains the potential of leakage. The elements and actors that are organized inside the frame also constitute channels that can lead to overflow. For the frame, in this case, the UNFCCC (and the Kyoto Protocol), to operate and to structure certain actions, it necessarily has to spread and circulate documents, scientific research, strategies, decisions, etc., on the internet, through meetings, summits, and organizational bodies. This is the only way the frame can be productive, but it also means that the climate framework is being adapted and performed in different and heterogeneous locations, which can lead to overflow, as in the case of the ITT Initiative. A former member of the Administrative Committee commented on this situation:

> My dream was to create the Yasuní Protocol, a different protocol, because the Kyoto Protocol is not the only instrument that may exist. The UNFCCC does not limit you to a single instrument. It says create instruments. Kyoto was created, so why not create the Yasuní Protocol or any other that addresses new ways of treating the carbon emissions issue.

The Yasuní Initiative worked upon the limitations of the Kyoto Protocol by proposing an alternative mechanism that involved "the active participation of developing countries in the mitigation of climate change" (Larrea et al. 2009: 5). By reframing the issue as primarily an extraction and not an emissions problem and replacing the concept of reduced emissions with *net avoided emissions*, the Yasuní proposal showed that the problem starts long before GHG emissions reach the atmosphere. Hence, climate change mitigation cannot only be concerned with what happens after fossil fuels have been extracted and burned; it also has to include extraction itself (Princen et al. 2015). As an initiative that clearly overflows the Kyoto Protocol, Yasuní-ITT points towards some of the current limitations produced by carbon as the translation unit, which overlooks many of the local complexities and entanglements. As a "boundary-object" (Star and Griesemer 1989), carbon allows "the framing and stabilization of actions, while simultaneously providing an opening on to other worlds, thus constituting leakage points where overflowing can occur" (Callon 1998b: 18).

By performing carbon translations calibrated to include what happens at the level of the underground, the Yasuní thesis connects climate change mitigation with the non-production of oil, which is an absent and politically invisible element in existing global climate strategies. Hence, focus is placed on the subsoil as a neglected geopolitical space (Valdivia 2015) in current climate negotiations. The following comment by a former Pachamama Foundation representative emphasizes how the Yasuní-ITT Initiative comprises an attempt to reframe the issue:

> I think that the most substantial and organizational contribution was that if you look at all the international negotiations, they always talk about emissions reduction. Climate change is a case of the greenhouse effect, which requires reducing emissions. That is the logic. For the first time, a small country, Ecuador, comes along saying it is not an emissions reduction issue. It is about going to the source, oil. It is about not emitting greenhouse gasses. Everything is reduction; the whole mitigation thing is about reduction. The fundamental debate is saying the problem is not this thing we cannot see, CO_2 "floating" in the atmosphere and nobody knows who created it, nobody knows why it is there and nobody takes ownership for it. The problem is that you take your car and you put gasoline in it and your car burns it. So, the problem is fossil fuels.

When the climate crisis is reframed as an extraction problem, which shifts the focus from carbon to fossil fuels (Princen et al. 2015), the solution is also reframed as a matter of avoiding CO_2 emissions, not only reducing them. Current climate solutions represent huge economic and environmental costs as new carbon reserves are constantly being released. Accordingly, end-of-pipe solutions are insufficient to combat the problem (Princen et al. 2015). In the case of Amazonia, climate change is entangled in fossil-fuels exploitation, loss of biodiversity, and complex territorialities, all interrelated issues that the Yasuní-ITT Initiative attempted to solve in a rather syncretic way by combining and simultaneously questioning elements from REDD and the Kyoto Framework. This overflow resulted in a precarious assemblage that struggled to link the atmosphere to the underground.

Yasuní: A national "love story" gone wrong?

> I think the strength of the initiative was that it was a complete project. It was complete because it had to do with climate change, indigenous peoples, biodiversity, renewable energy; well, I was very much in love with the project.[14]

On August 15, 2013, President Correa announced on national TV the decision to terminate the Yasuní-ITT Initiative, as it had not achieved the expected results: 3.6 billion dollars over a 13-year period. After more than six years of campaigning, with several obstacles and even participants who withdrew from their initial commitment (the case of Germany), only 13.3 million dollars had actually been deposited in the ITT Trust Fund, which comprised 0.37% of the expected amount (Presidencia de la República 2013). That the Yasuní Initiative had managed to position itself as a real alternative, especially among young people, was something that the President was highly aware of, and several times during his intervention, he directly addressed the country's young population. According to President Correa, the Yasuní-ITT Initiative had been an opportunity for the international community to move from the heights of rhetoric to real action, demanding co-responsibility in combating climate change (Presidencia de la República 2013). Sadly, "the world has failed us", the President announced. Consequently, the government had decided to request authorization from the National Assembly to open up the ITT block for oil drilling to fulfill the goals of the National Plan for Good Living.[15] This movement from plan A to plan B constituted a shift in the government's environmental policies that generated a new moment of translation, as the Yasuní-ITT proposal returned to civil society where it originated. An environmentalist involved commented about this situation:

> I believe the date was very important because by then, the state had lost much of the initiative's credibility after having Mrs. Ivonne Baki (Secretary of State for the Yasuní Initiative) traveling around the world. People from civil society did not even have a connection. They no longer cared since there were inconsistencies. I think the date was important, as I told you, especially young people had heard about the initiative for six years and we were convinced it was the best thing ever. So, it allowed the initiative to return to civil society, and it was with Yasunidos.

During the time it lasted, the Yasuní-ITT Initiative had a profound impact on young people in Ecuador, as it represented a different way to engage with nature than what had been the case since Ecuador became an oil producer in the 1970s. Consequently, many people saw the initiative as an opportunity to "defend life" or, in other words, to defend the interrelatedness of humans and nonhumans in the Amazon. Yasunidos initially articulated around key points of the Yasuní-ITT Initiative, namely defending the rights of indigenous peoples living in isolation and of nature to preserve the rich cultural and biological diversity that characterizes the region. Overall, the ITT Initiative constituted an "environmental awakening" among a considerable group of urban, middle-class youth, who decided to form a collective to continue defending non-exploitation of the ITT field when the National Assembly started the process towards declaring blocks 31 and 43 (ITT) of national interest. Yasunidos can be understood, therefore, as the effect of network translations that traveled from civil society to government institutions and back to civil society again. With several new ideas and proposals, civil society organizations attempted to halt the opening of the ITT field. One of the most clearly articulated proposals was plan C, which was a direct response to the government's plan B, as it proposed an alternative mechanism to obtain the financial resources without exploiting the ITT field. A representative from CDES commented on the political movement of the debate from nature conservation to social equality:

The discourse generated a false dichotomy between more oil or more poverty as if the development issue was a matter of economic growth. So, what we said then was, let's see, even if we do not agree with what the government is proposing, it has just placed the debate where it belongs, which means the debate was appropriate. The conservation issue is also a matter of social equality but they are presenting a false dichotomy. So, we decided to draw the economic issue into the debate. Until then, that was something nobody talked about. Well, "plan C" positioned it and Yasunidos adopted it as their alternative. Well, not officially but at least they repeated it constantly.

Since the government increasingly performed Yasuní as an economic issue tied to possibilities of development, plan C tried to show that there were other ways to obtain these resources, for example, by adding 1.5% in taxes to the 110 economic groups that have benefited the most from economic growth and stability over the last period (Iturralde 2013: 21). The Yasunidos collective largely embraced this economic argument and translated it into its communications strategy to argue that it was not a question of conservation of nature versus solving the development issue, as it was possible to do both, even in the absence of international contributions.

As a new network started to emerge, Yasuní went from being translated principally as a matter of defending the rights of humans and nonhumans to becoming a matter of testing the political system. Consequently, the new *problematization* process set out to explore the boundaries of democracy and the citizens' rights to participate directly in political decision-making. The constituent process and Assembly (2007–2008) had been something close to a "legal laboratory" that produced a new constitution with a series of legal novelties: Plurinationality, the Rights of Nature (Pachamama), a series of collective and territorial rights, etc. The Yasunidos collective decided to test the Constitution's true agency by setting the second part of art. 407 to work: "Exceptionally, these resources can be tapped at the substantiated request of the President of the Republic and after a declaration of national interest issued by the National Assembly, which can, if it deems it advisable, convene a referendum".

In order to convene a national referendum on whether the state should or should not exploit the petroleum reserves in the ITT, Yasunidos had to collect 584,116 signatures from citizens (5% of the number of registered voters in the last elections) who supported a national referendum on Yasuní-ITT (El Telégrafo 2013). With a different conception of politics and a horizontal structure that differed from the traditional political organizations in Ecuador, Yasunidos developed into a new way of seeking participation in national issues by testing the Constitution in practice. In the words of a member of the collective, "by taking the constitutional content and bringing it to life".

With huge optimism and creativity, Yasunidos started their campaign to enroll and mobilize the country for a national referendum on oil extraction in Yasuní's rainforest. This was the first time in the country's history that a request for a national referendum originated in civil society. The referendum was never carried out, however, since the National Electoral Council found only 359,761 signatures to be valid of around 750,000 signatures presented by Yasunidos, alleging that several signatures were repeated and some were fictitious (El Telégrafo 2014). This result was contested by representatives of the Yasunidos collective. Nevertheless, Yasunidos managed to place nature and the environmental issue at the center of national political life. The process can be seen in the light of what Escobar (1999) calls a "politics of hybrid natures". In this kind of politics, organic nature becomes an important anchor for processes of resistance and local struggle (Escobar 1999: 13); however, capitalist nature is also present. Whereas the Yasunidos attempted to disconnect the territory from the oil industry, they also emphasized the importance of alternative translocal economic strategies for the Amazon region, such as tourism, bio-knowledge, and "eco-experiences" in which capitalist and organic natures largely overlap. Still, the ITT issue acquired a broader scope with Yasunidos, as it was translated as a site of contestation that extended beyond the Amazon and the socioenvironmental conflicts of the rainforest. By articulating Yasuní as a matter of democratic practice, it simultaneously became a site for exploring both the Constitution's agency and the boundaries of the current political system.

Notes

1   Translation by the author.
2   All quotations from the Management Plan for Yasuní National Park issued by the Ministry of the Environment (Ministerio del Ambiente 2011) are my own translations.
3   Presidential decree 2187, January 3, 2007 (Fontaine 2007b: 84).
4   In 2006, the Inter-American Commission on Human Rights granted precautionary measures in favor of the Tagaeri and the Taromenane. The Commission requested the Ecuadorian state to adopt measures after receiving information about the killing of members of the Taromenane group during reprisals related to illegal logging. The main objective of these precautionary measures was to protect the lives and the integrity of uncontacted peoples in the Ecuadorian Amazon as well as their territories by preventing third parties from entering the area.
    Inter-American Commission on Human Rights (2006): http://www.cidh.org/medi das/2006.eng.htm.
5   Interview with a representative of CDES, Quito, February 24, 2015.
6   The audience of Eduardo Pichilingue (ex-coordinator of the Plan for Precautionary Measures) in the National Assembly's Commission on Biodiversity and Natural Resources, September 12, 2013: https://www.youtube.com/watch?v=E4Vtzuczrrw.
7   Alianza País was the governing party under the presidency of Rafael Correa (2007–2017). In 2017, Lenín Moreno, vice-president during Correa's first presidential term, won the elections. The conflict and rupture with his predecessor fractioned Alianza País, as those who sympathized with Correa left the party accusing Moreno of treason.
8   Translation by the author.
9   Speech of President Rafael Correa, High Level Dialogue on Climate Change of the 62nd Period of Sessions of the General Assembly of the United Nations, New York, September 24, 2007: http://mptf.undp.org/yasuni.
10   UNDP Multi-Partner Trust Fund Office: Yasuní-ITT Fact Sheet: http://mptf.undp .org/yasuni.
11   The first period of commitment required industrialized countries to reduce their emissions by 5% against 1990 levels. During the second commitment period, the intention was to reduce emissions at least 18% below 1990 levels. UNFCCC (n.d.): https://unfccc.int/kyoto_protocol.

12 UNFCCC: Emissions Trading: https://unfccc.int/process/the-kyoto-protocol/mech anisms/emissions-trading.
13 Using field inventory data, the authors compared Yasuní's diversity to other sites in western Amazonia and globally. Data suggest that Yasuní "is among the most biodiverse places on Earth, with apparent world richness records for amphibians, reptiles, bats, and trees" (Bass et al. 2010: 1).
14 Interview with a member of the Yasunidos Collective, Quito, April 22, 2015.
15 This national planning document can be seen as an interpretation of the principle of *Sumak Kawsay*, which is explained in the following terms:

> Good Living, or Sumak Kawsay, is a mobilizing concept that offers alternatives to mankind's problems. Sumak Kawsay strengthens social cohesion, community values and encourages the active involvement of individuals and collectives in major decision making in order to construct their destiny and happiness.
> (National Secretariat for Planning and Development (Senplades) 2013: 21)

References

Acción Ecológica. (2000, August 20). *Campaña Amazonía por la vida*. Retrieved from https://www.accionecologica.org/amazonia-por-la-vida/.

Álvarez, K. (2016). Lugares cargados de memoria: Aproximaciones hipotéticas sobre la construcción de identidad y territorio en los Tageiri y Taromenane. *Antropología Cuadernos de Investigación*, 16, 69–84.

Álvarez, Y. (2013). Una propuesta desde el ecologismo para proteger a una parte del Parque Nacional Yasuní. In I. Narváez, M. De Marchi, & S. E. Pappalardo (eds), *Yasuní Zona de Sacrificio: Análisis de la Iniciativa ITT y los Derechos Colectivos Indígenas* (pp. 80–101). Quito: Flacso.

Arsel, M. (2012). Between Marx and markets? The state, the "left turn" and nature in Ecuador. *Tijdschrift voor Economische en Sociale Geografie*, 103 (2), 150–163.

Asamblea Nacional. (2013, September 30). *Informe para el segundo debate en el pleno de la Asamblea Nacional sobre proyecto de resolución de declaratoria de interés nacional para la explotación de los bloques 31 y 43 dentro del Parque Nacional Yasuní*. Retrieved from https://www.eltelegr afo.com.ec/images/eltelegrafo/Actualidad/2013/02-10-13-informe-segundo-debate-yasuni.pdf.

Asdal, K. (2008a). Enacting things through numbers: Taking nature into account/ing. *Geoforum*, 39 (1), 123–132.

Asdal, K. (2008b). On Politics and the little tools of democracy: A down-to-earth approach. *Distinktion: Scandinavian Journal of Social Theory*, 9 (1), 11–26.

Bartlett, L., & Vavrus, F. (2017). *Rethinking Case Study Research: A Comparative Approach*. New York: Routledge.

Bass, M. S., Finer, M., Jenkins, C. N., Kreft, H., Cisneros-Heredia, D. F., & McCracken, S. F. (2010). *Global conservation significance of Ecuador's Yasuní National Park*. Retrieved from Plos One 5 (1): http://journals.plos.org/plosone/article?id=10.1371/journal.pone.00 08767.

Blok, A., & Jensen, T. E. (2011). *Bruno Latour: Hybrid Thoughts in a Hybrid World*. London: Routledge.

Bridge, G. (2011). Resource geographies I: Making carbon economies, old and new. *Progress in Human Geography*, 35 (6), 820–834.

Callon, M. (1986). Some elements of a sociology of translation: Domestication of the scallops and the fishermen of St. Brieuc Bay. In J. Law (ed), *Power, Action and Belief: A New Sociology of Knowledge?* (pp. 196–233). London: Routledge & Kegan Paul.

Callon, M. (1998a). An essay on framing and overflowing: Economic externalities revisited by sociology. In M. Callon (ed), *The Laws of the Markets* (pp. 244–269). Oxford: Blackwell Publishers.

Callon, M. (1998b). Introduction: The embeddedness of economic markets in economics. In M. Callon (ed), *The Laws of the Markets* (pp. 1–57). Oxford: Blackwell Publishers.

Colleoni, P., & Proaño, J. (2010). *Caminantes de la Selva: Los pueblos en aislamiento en la Amazonía Ecuatoriana. Informe IWGIA 7.* Retrieved from https://www.iwgia.org/images/public ations//0275_Informe_7_eb.pdf.

Constitution of the Republic of Ecuador. (2008, October 20). Retrieved from Political Database of the Americas: http://pdba.georgetown.edu/Constitutions/Ecuador/engl ish08.html.

Correa, R. (2007, September 24). *Speech given at the High Level Dialogue on Climate Change of the 62 Period of Sessions of the General Assembly of the United Nations.* Retrieved from htt://mptf. undp.org/yasuni.

De la Cadena, M. (2010). Cosmopolitics in the Andes: Conceptual reflections beyond politics. *Cultural Anthropology*, 25 (2) 334–370.

El Comercio. (2016, February 17). *Fundación Labaka pide que se detenga la actividad extractiva en Armadillo, por la vida de los pueblos ocultos de Yasuní.* Retrieved from www.elcomercio .com: http://www.elcomercio.com/actualidad/fundacionlabaka-armadillo-petroleo-taromenane-yasuni.html.

El Telégrafo. (2013, August 19). *Consulta por el Yasuní requiere 584.116 firmas.* Retrieved from http://www.eltelegrafo.com.ec/noticias/economia/8/consulta-por-el-yasuni-requiere -584-116-firmas

El Telégrafo. (2014, May 7). *Yasunidos no alcanza firmas para consulta popular sobre Yasuní.* Retrieved from www.eltelegrafo.com.ec: http://www.eltelegrafo.com.ec/noticias/p olitica/2/yasunidos-no-alcanza-firmas-para-consulta-popular-sobre-yasuni.

Escobar, A. (1999). After nature: Steps to an antiessentialist political ecology. *Current Anthropology*, 40 (1), 1–30.

Escobar, A. (2001). Culture sits in places: Reflections on globalism and subaltern strategies of localization. *Political Geography*, 20 (2), 139–174.

Escobar, A. (2011). Sustainability: Design for the pluriverse. *Development*, 54 (2), 137–140.

Escobar, A. (2015). Territorios de diferencia: La ontología política de los "derechos al territorio". *Cuadernos de Antropología Social*, pp. 25–38.

Fontaine, G. (2007a). *El Precio del Petróleo: Conflictos Socio-Ambientales y Gobernabilidad en la Región Amazónica.* Quito: Flacso, IFEA, Abya-Yala.

Fontaine, G. (2007b). Problemas de la cooperación institucional: El caso del comité de gestión de la reserva de biosfera Yasuní. In G. Fontaine, & I. Narváez (eds), *Yasuní en el Siglo XXI: El Estado Ecuatoriano y la Conservación de la Amazonía* (pp. 75–127). Quito: Flacso.

Fontaine, G., & Narváez, I. (2007). Prólogo: Problemas de la gobernanza ambiental en el Ecuador. In G. Fontaine, & I. Narváez (eds), *Yasuní en el Siglo XXI: El Estado Ecuatoriano y la Conservación de la Amazonía* (pp. 13–31). Quito: Flacso.

Franco, J. C. (2013). Territorio waorani: Problemática y el proceso extractivo en el Yasuní. In I. Narváez, M. De Marchi, & S. E. Pappalardo (eds), *Yasuní Zona de Sacrificio: Análisis de la Iniciativa ITT y los Derechos Colectivos Indígenas* (pp. 141–173). Quito: Flacso.

Fundación Pachamama. (2007, November 29). *El Plan Verde.* Retrieved from http://pla nverdecuador.blogspot.no/2007/11/bienvenidos-al-plan-verde.html

Gallardo Fierro, L. (2016). Oil or "life": The dilemma inherent in the Yasuní-ITT Initiative. *Extractive Industries and Society*, 3 (4), 939–946.

Inter-American Commission on Human Rights. (2006). *Precautionary measures granted by the IACHR in 2006.* Retrieved from www.cidh.org: http://www.cidh.org/medidas/2006 .eng.htm.

International Labour Organization. (1989). *Indigenous and Tribal Peoples Convention (No. 169).* Retrieved from http://www.ilo.org/dyn/normlex/en/f?p=NORMLEXPUB:12100 :0::NO::P12100_INSTRUMENT_ID:312314.

Iturralde, P. J. (2013). *Plan C: Redistribución de la riqueza para no explotar el Yasuní y salvaguardar los indígenas aislados.* Quito: CDES, Observatorio de Derechos Colectivos del Ecuador.

Lalander, R. (2014). The Ecuadorian resource dilemma: Sumak Kawsay or development? *Critical Sociology,* 42 (4–5), 623–642.

Larrea, C., & Warnars, L. (2009). Ecuador's Yasuní-ITT initiative: Avoiding emissions by keeping petroleum underground. *Energy for Sustainable Development,* 13 (3), 219–223.

Larrea, C., Greene, N., Rival, L., Sevilla, E., & Warnars, L. (2009). *Yasuni-ITT Initiative: A big idea from a small country.* Retrieved from http://mptf.undp.org/document/download/ 4545.

Larsen, P. B. (2017). Oil territorialities, social life, and legitimacy in the Peruvian Amazon. *Economic Anthropology,* 4 (1), 50–64.

Latour, B. (1987). *Science in Action: How to Follow Scientists and Engineers Through Society.* Cambridge, MA: Harvard University Press.

Latour, B. (1993). *We Have Never Been Modern.* Cambridge, MA: Harvard University Press.

Latour, B. (2004). *Politics of Nature: How to Bring the Sciences into Democracy.* Cambridge, MA: Harvard University Press.

Latour, B. (2005). *Reassembling the Social: An Introduction to Actor-Network Theory.* New York: Oxford University Press.

Latour, B. (2007). To modernize or to ecologize? That is the question. In K. Asdal, B. Brenna, & I. Moser, *Technoscience: The Politics of Interventions* (pp. 249–272). Oslo: Unipub.

Law, J. (2003). *Traduction/Trahison: Notes on ANT.* Retrieved from Centre for Science Studies, Lancaster University: http://www.comp.lancs.ac.uk/sociology/papers/Law-Traduction-Trahison.pdf.

Law, J., Afdal, G., Asdal, K., Lin, W.-y., Moser, I., & Singleton, V. (2013). Modes of syncretism: Notes on noncoherence. *Common Knowledge,* 20 (1), 172–192.

Martin, P. L. (2015). Leaving oil under the Amazon: The Yasuní-ITT Initiative as a postpetroleum model? In T. Princen, J. P. Manno, & P. L. Martin (eds), *Ending the Fossil Fuel Era* (pp. 119–144). Cambridge, MA: MIT Press.

Martínez, E. (2009). *Yasuní: El Tortuoso Camino de Kioto a Quito.* Quito: Abya-Yala.

Martz, J. D. (1987). *Politics and Petroleum in Ecuador.* New Brunswick: Transaction Books.

Ministerio de Justicia, Derechos Humanos y Cultos. (2013a, August 22). *Informe sobre posibles señales de presencia de pueblos indígenas aislados en los bloques 31 y 43 (ITT).* Retrieved from http://www.geoyasuni.org/wp-content/uploads/2013/09/All3MJDHC.pdf.

Ministerio de Justicia, Derechos Humanos y Cultos. (2013b, November). *Proyecto "Implementación de Estación de Monitoreo Shiripuno".* Retrieved from http://www.justicia. gob.ec/wp-content/uploads/2016/05/PROYECTO-SHIRIPUNO.pdf

Ministerio del Ambiente. (2011). *Plan de Manejo del Parque Nacional Yasuní.* Quito, Ecuador. Retrieved from http://suia.ambiente.gob.ec/documents/10179/242256/45+PLAN +DE+MANEJO+YASUNI.pdf/8da03f55-1880-4704-800e-d5167c80089c.

Mol, A., & Law, J. (2002). Complexities: An introduction. In J. Law, & A. Mol (eds), *Complexities: Social Studies of Knowledge Practices* (pp. 1–22). Durham: Duke University Press.

Müller, M. (2015). Assemblages and actor-networks: Rethinking socio-material power, politics and space. *Geography Compass,* 9 (1), 27–41.

Narváez, I. (1996). *Huaorani-Maxus: Poder Étnico vs. Poder Transnacional*. Quito: Fundación Ecuatoriana de Estudios Sociales (FESO).

Narváez, I., De Marchi, M., & Pappalardo, E. (2013). Prólogo Yasuní: En clave de derechos y como ícono de la transición, para ubicarse en la selva de proyectos. In I. Narváez, M. De Marchi, & S. E. Pappalardo (eds), *Yasuní Zona de Sacrificio: Análisis de la Iniciativa ITT y los Derechos Colectivos Indígenas* (pp. 9–26). Quito: Flacso.

National Secretariat of Planning and Development. (2013). *National Plan for Good Living*. Retrieved from http://www.buenvivir.gob.ec/versiones-plan-nacional;jsessionid=16 5A89C0170390FF6834DC69D481F493.

Pappalardo, S. E., De Marchi, M., & Ferrarese, F. (2013). Uncontacted Waorani in the Yasuní Biosphere Reserve: Geographical validation of Zona Intangible Tagaeri Taromenane (ZITT). *Plos One*, 8 (6) Retrieved from https://doi.org/10.1371/journal. pone.0066293.

Pellegrini, L., Arsel, M., Falconí, F., & Muradian, R. (2014). The demise of a new conservation and development policy? Exploring the tensions of the Yasuní ITT initiative. *Extractive Industries and Society*, 1 (2), 284–291.

Pigrau, A. (2012). *The Texaco-Chevron Case in Ecuador. EJOLT Factsheet No. 42*. Retrieved from http://www.ejolt.org/wordpress/wp-content/uploads/2015/08/FS-42.pdf.

Presidencia de la República, Ecuador. (2013, August 15). *Anuncio a la Nación Iniciativa Yasuní-ITT*. Retrieved from http://www.presidencia.gob.ec/wp-content/uploads/downloa ds/2013/08/2013-08-15-AnuncioYasuni.pdf.

Princen, T., Manno, J. P., & Martin, P. L. (2015). The problem. In T. Princen, J. P. Manno, & P. L. Martin (eds), *Ending the Fossil Fuel Era* (pp. 3–36). Cambridge MA: MIT Press.

Rival, L. M. (2015). *Transformaciones Huaoranis: Frontera, Cultura y Tensión*. Quito: Universidad Andina Simón Bolívar, Latin American Centre of Oxford, Abya-Yala.

Rivas Toledo, A., & Lara Ponce, R. (2001). *Conservación y Petróleo en la Amazonía Ecuatoriana: Un Acercamiento al Caso Huaorani*. Quito: EcoCiencia, Abya-Yala.

Secretaría de Hidrocarburos del Ecuador (2015). *Mapa de bloques petroleros del Ecuador continental*. Retrieved from http://www.secretariahidrocarburos.gob.ec/mapa-de-bloqu es-petroleros/Mapa- Bloques-Petroleros-actualizado-en-la-WEB-25_09_2015.

Sovacool, B. K., & Scarpaci, J. (2016). Energy justice and the contested petroleum politics of stranded assets: Policy insights from the Yasuní-ITT Initiative in Ecuador. *Energy Policy*, 95, 158–171.

Star, S. L., & Griesemer, J. R. (1989). Institutional ecology, "translations" and boundary objects: Amateurs and professionals in Berkeley's museum of vertebrate zoology. *Social Studies of Science*, 19 (3), 387–420.

Strathern, M. (2004). *Partial Connections*. Walnut Creek: Roman & Littlefield Publishers.

UNDP Multi-Partner Trust Fund Office. (n.d.). *Ecuador Yasuní ITT Trust Fund Fact Sheet*. Retrieved from http://mptf.undp.org/yasuni/en.

United Nations Framework Convention on Climate Change. (n.d.). *What is the Kyoto Protocol?* Retrieved from https://unfccc.int/kyoto_protocol.

Valdivia, G. (2015). Oil frictions and the subterranean geopolitics of energy regionalism. *Environment and Planning A: Economy and Space*, 47 (7), 1422–1439.

Yasuní ITT Trust Fund. (2011, September). *Operating Procedures of the Yasuní ITT Trust Fund Steering Committee*. Retrieved from http://mptf.undp.org/yasuni.

6   Lofoten and Yasuní-ITT

Comparing interfering networks

After a brief introduction on oil and climate as epistemic objects, which serves as a frame for this chapter, here I will draw the cases together by comparing and contrasting key elements and themes from the previous case analyses presented in Chapters 2–5. As a multi-sited and multi-level inquiry, the focus of comparison will move between the cases but also between scales: The controversies linked to Lofoten and Yasuní are translocal political processes that constitute spatially dispersed fields (Bartlett and Vavrus 2017: 43), which implies tracing relationships across local, national, and global levels. Within a process-oriented comparative case study, scales are conceptualized not as fixed or settled dimensions but as effects of political, economic, environmental, and technological processes. These have, on some occasions, expanded and distributed networks through dense and interconnected links, whereas on other occasions, relations and linkages have not managed to travel and expand across space and time. In other words, their materials and configurations have been less durable. The comparison that follows primarily emphasizes examining similarities and differences across sites (horizontal axis) by focusing on national and international policies and frameworks (vertical axis). Hence, the chapter constitutes a cross-case analysis (Creswell 2013) on how various encounters between oil and climate influence aspects such as territorialization and spatial ordering, the framing of climate change, and the production of natures and socioenvironmental risks.

Oil and climate as intersecting epistemic objects

In order to approach oil and climate as interrelated epistemic objects, I again draw on Knorr Cetina who defines these objects "by their lack of completeness of being and their nonidentity with themselves" (2001: 176). As knowledge objects, oil and climate have progressively acquired new identities due to a series of connections presented in research reports, journal articles, modeled trajectories, and calculated predictions provided by national and international organizations and entities. Scientific knowledge about the positive correlation between fossil-fuel combustion and climate change was available several decades before the Brundtland Commission presented its report *Our Common Future* in 1987.

While there had been a growing consensus within the scientific community, the document added *political facticity* to the thesis of anthropogenic global warming. The report specifically emphasized that the world was approaching the limits of nature's systems, a situation provoked by intensified tapping and use of natural resources that could threaten future life on Earth. Regarding the link between fossil fuels and climate change, the report stated:

> The "greenhouse effect", one such threat to life-support systems, springs directly from increased resource use. The burning of fossil fuels and the cutting and burning of forests release carbon dioxide (CO_2). The accumulation in the atmosphere of CO_2 and certain other gases traps solar radiation near the Earth's surface, causing global warming. This could cause sea level rises over the next 45 years large enough to inundate many low-lying coastal cities and river deltas. It could also drastically upset national and international agricultural production and trade systems.
>
> (World Commission on Environment and Development 1987: 6–7)

On the one side, and according to the text, it is clear that the burning of fossil fuels and deforestation is changing the climate by contributing to higher levels of CO_2 in the atmosphere. On the other side, this interrelationship is also *politically doing something* to petroleum resources, as the climate issue is *endangering them as future economic assets* in a world that requires a transition towards a low-carbon energy system. Ever since oil became an important energy commodity at the end of the 19th century, thanks to the invention of the internal combustion engine (Bradshaw 2014: 5), oil has been studied, calculated, and mapped to establish the location and volumes of petroleum reservoirs. Research has also led to new ways of surveying underground geological structures, accompanied by the corresponding development of drilling technology. Furthermore, a large number of oil analysts around the world are dedicated to studying everything from market tendencies and fluctuating oil prices to oil companies and their assets. Studies on energy security and the geographical distribution of oil reserves are also adding knowledge to the petroleum field. As previously unknown characteristics and connections have surfaced, new uncertainties have generated further research and lines of inquiry. Consequently, it is oil's unfolding, complex, and ambiguous identity, in other words its *unfinishedness* that makes it a "meaning-producing and practice-generating" object (Knorr Cetina 2001: 183). However, the highly unstable and unsettled character of epistemic objects makes it more appropriate to think of them as "processes and projections rather than definite things" (Knorr Cetina 2001: 181). This is not just the case for oil; climate also displays a similar capacity to unfold continuously in different directions and research fields. As an epistemic object, climate has been assembled in various studies on long-term weather patterns, ocean–atmosphere dynamics, ice processes, climate modeling, and weather statistics. Researchers have tried to understand the role and behavior of the Gulf Stream and phenomena such as El Niño and La Niña[1]. Multiple studies related to critical concentrations of greenhouse gases (GHGs) in the atmosphere, the greenhouse effect, and the carbon cycle have generated attempts to set up carbon budgets and calculate global thresholds. Today, climate is an epistemic object that is being assembled through various intersecting disciplines such as climatology, oceanography, and biogeochemistry. The various encounters between oil and climate have also generated a new web of uncertainties across scales and sites that translate into increased scientific activity, "green" engineering, and development of renewable energy sources. Although knowledge objects have never really been things in themselves due to their relational character, oil and climate have become increasingly complex, open, and unsettled as they cross paths. This "lack in completeness of being" multiplies in oil and climate's mutual encounters and can, therefore, "be conceived of as unfolding structures of absences" (Knorr Cetina 2001: 182; emphasis removed).

Furthermore, oil has been politically modified and reassembled by climate change. Conversely, oil is also participating and contributing to global warming, the very same process that is causing its own transformation as a knowledge object. Consequently, epistemic objects are always in the process of becoming, transitioning towards what they have yet to become. Regarding their *unfolding ontology*, Knorr Cetina explains that epistemic objects can have multiple instantiations (figurative, mathematical, material, etc.), which are always partial. These "partial objects" are not just stand-ins for the real knowledge object but are manifestations of its unfolding ontology (2001: 182). Consequently, Knorr Cetina argues that naming is a way to hold things together: A stable name "folds up" a sequence of unfolding objects as it translates between places and times (Knorr Cetina 2001: 184). Oil and climate are, therefore, not definite things, as they are always mutating in new, uncertain network translations. They intersect and meet, but not always, so we are back to partial connections, in other words. The following sections of the chapter will analyze how these partial connections emerge and work across scales and, thereby, unite the cases of Lofoten and Yasuní-ITT (Ishpingo, Tambococha, and Tiputini).

The inscription of territorial conflicts

I have examined the territorial nature of petroleum activities in the foregoing chapters and, more specifically, how territorializations are "enacted into being" (Law 2007: 12) by policies, maps, legal frameworks, concessions, management plans, etc. These processes are highly translocal. Therefore, I want to emphasize the importance of moving horizontally and vertically (sometimes we even need to calibrate our scope to the level of the underground) in order to trace "spatially non-contiguous assemblages of human and nonhuman actors" (Bartlett and Vavrus 2017: 77). To compare how Lofoten and Yasuní are enacted as territories and how territorial conflicts are politically handled, we need to get close up on sociomaterial practices concerned with territorialization and place-making (Escobar 2001). Comparing is also the action of juxtaposing or bringing together, which provides an opportunity to look for links, routes, dispatchers, or fine threads that at some point entangle or intertwine the cases. Both in Norway and in Ecuador, the oil industry is pushing for access to new areas that are classified as valuable and vulnerable and as protected areas in the countries' management plans. The oil industry demands space to maintain current levels of exploration and extraction. Without a political framework that guarantees the possibility of accessing and opening new fields, the oil industry argues that it cannot sustain long-term planning or investments in the hydrocarbon sector. Combined with declining production in several mature oilfields in the North Sea and in the northern Ecuadorian Amazon, resource availability is presented in both cases as a pressing matter. The ongoing political processes related to Lofoten and Yasuní should, therefore, be seen in conjunction with oil frontiers that are currently expanding and "on the move". Hence, the Lofoten, Vesterålen, and Senja (LoVeSe) and Yasuní-ITT controversies are effects of several translations that strategically work to enroll new territories to keep the networks "up and going" as space constitutes vital infrastructure for the petroleum industry. Territories are made accountable to the networks through enrollment; nevertheless, oil resources are always surrounded by a considerable degree of uncertainty. "The level of proven reserves is neither stable nor absolute" (Bradshaw 2014: 26), as it depends on economic and technical viability, which makes it a rather *fluid* category. Some areas can be accessible from a legal point of view, but conditions such as low oil prices, or costly technological solutions, affect the possibility at any given moment of a project being developed and coming on-stream. Consequently, resource availability and access are prerequisites for maintaining optimal levels of productivity and stable oil exports.

The way these circumstances shape the national debates regarding the possibility of drilling in environmentally sensitive areas, presents several similarities between the cases. In Norway, the arguments employed to legitimize and politically anchor the possibility of petroleum operations in areas that have been classified as vulnerable in research documents and management plans are tied to the importance of preserving the country's role as a stable international energy producer. Many sectors within Norwegian society believe this is fundamental if the country wants to preserve the welfare state in its current form. A reorientation or a change of course in Norway's oil policies could, therefore, threaten the very existence of the welfare state and the Norwegian model. This preoccupation is present in the following reflection provided by a local Conservative Party politician in Lofoten:

> It is a very important field because if we close down this side of the Norwegian industry and do not open more fields and do not continue exploring, it will have dramatic consequences for Norway as a nation, for the national economy in a long-term perspective, and it will increase unemployment, you name it. When we know how oil dependent the Norwegian economy is, then we depend on … the industry depends on having a stable flow of new areas and all those things, that they have progression in both exploration programs and extraction to make long-term investments within the sector.

In a similar fashion, the oil lobby has repeatedly emphasized the importance of providing stable conditions for the industry as a way to secure the state's income and its ability to finance welfare in the future. In an interview with E24 (e-journal for economic news), the former Director General of the Norwegian Oil and Gas Association, Karl Eirik Schjøtt-Pedersen, emphasized the role of the country's largest political forces, the Conservative Party (*Høyre*) and the Labour Party (*Arbeiderpartiet*), in securing the welfare state by providing their political support in favor of maintaining a high activity level in the petroleum and gas sector. According to Schjøtt-Pedersen, the best way to achieve this objective is by opening up new exploration areas for the industry and, thereby, also guaranteeing high activity in 2030 and 2040 (E24 2017). The strong focus on welfare and the developmental possibilities that come with oil is one of the key elements in *oil as safety*, an ordering mode that enacts exploitation of petroleum resources as an economic safety net at present and for future generations. However, as an organizing economic agent, *oil cannot achieve development by itself* as it requires interinstitutional links, a strong legal framework that governs the sector, and political accountability. These elements form the web of relations that, over time, have produced the Norwegian way of managing petroleum resources or, in short, the Norwegian way of "doing oil". This particular approach is shared with developing countries that are interested in learning from Norway's oil experience through *Oil for Development* (OfD), a program established in 2005 by the Norwegian Agency for Development Cooperation. The program's objective is to advise other countries on how oil can become a developmental agent, as it has often proven difficult in many countries to translate oil into welfare for the population and, thereby, escape the resource curse (Norwegian Agency for Development Cooperation; NORAD 2013). Storylines about oil as the engine behind economic growth and welfare are many and strong in Norway. Naturally, they exist within sectors that wish to extend the petroleum age and continue with business as usual, but we also find them among stakeholders who want to slow down Norway's petroleum production due to climate change, and transit towards a less oil-dependent economy. One of the main challenges is how to move away from something which has been a national success. This is where *oil as safety* transits towards *oil as insecurity*. Here, climate science plays an important role as it disrupts and interferes with former petroleum policies. The oil-fueled economy could be at a crossroads, as it is increasingly being threatened by climate change. It is, therefore, important to take into account the active role that Norway has taken on in several parts of the world by advising developing countries (upon request) on how to achieve a sustainable oil economy by sharing some of the lessons learned from Norway's own petroleum experience.

When Erik Solheim, the former Norwegian Minister of the Environment and International Development, met in Quito with members of the Ecuadorian government in 2007 to talk about petroleum management (El Diario.ec 2007), the recently elected president had started a process of rearranging Ecuador's petroleum sector to gain increased national control over the resources. While the OfD program was part of the talks, Ecuador also wanted to discuss the recent international launch of the Yasuní Initiative and the possibilities for financial support to keep the oil in the ground (Klassekampen 2007). As we know, Norway declined to participate, as the government feared the possibility of carbon leakage, since

Ecuador could, in theory, intensify oil extraction in other parts of the Amazon outside the ITT block. Whereas the OfD program links development to a series of institutional arrangements and legal regulatory frameworks that have to be in place for oil to generate growth and sustainable development, a basic condition for this approach is exploiting petroleum resources. Ecuador, conversely, wanted to establish a different model based on development from oil reserves that were to be kept in the ground or, in other words, development based on non-extraction of petroleum resources and contributions from the international community to finance renewable energy, technological innovation, and social development programs. The logics underpinning the two approaches can, therefore, be seen as conflicting. Later in this chapter, I will discuss similarities and differences between Norwegian and Ecuadorian climate proposals and how they specifically intersect at different levels. Here, I only want to emphasize the following: Whereas Ecuador's Yasuní-ITT Initiative problematized oil extraction as it directly threatens the forest and its peoples, besides releasing hydrocarbons that initiate their journey towards the atmosphere and, thereby, heat the planet, the Norwegian proposals and programs tend to keep these issues apart despite their growing glocal entanglements. A premise for OfD is oil extraction, whereas the Yasuní-ITT Initiative can be seen as part of an emerging trend of "transition discourses", which, according to Escobar often "are keyed in to the need to move to post-fossil fuel economies" (2011: 138). Although Ecuador is currently not enlisted among the countries participating in the OfD program, Norway's emphasis and promotion of poverty reduction through "economically, environmentally and socially responsible management of petroleum resources which safeguards the needs of future generations" (NORAD 2011) has, nevertheless, traveled around the globe. Many countries have looked to Norway and Canada for examples of sustainable management of petroleum and mineral resources. President Correa enrolled these ideas when the decision to terminate the ITT Initiative was broadcasted on national TV in August 2013. In response to those sectors and stakeholders who claimed that oil extraction was incompatible with preserving Yasuní, the president replied:

> They have fooled us with a false dilemma: Everything or nothing; exploit the ITT or the survival of Yasuní. This false dilemma is part of an even bigger false dilemma: Nature or extractivism. Norway is an oil-producing country, and it is one [of the countries] that best protects its nature (besides being a champion in human development). Canada is a mining country, and it has the biggest reserves of fresh water on the planet.
>
> As a consequence, the real dilemma is: 100% of Yasuní and no resources to satisfy the urgent necessities of our people or 99.9 % of Yasuní intact and close to 18 billion dollars to overcome poverty especially in Amazonia, paradoxically the region with the highest incidence of poverty[2].

Ecuador had worked for several years to internationally position its own development and climate model. However, faced with low financial support, the

Ecuadorian government reassembled the dilemma between conservation and use of natural resources. In the speech, there is clearly a step aside from public policies that took into consideration both the rights of humans and the rights of nature. By altering the previous dynamics between the underground and the surface, the ITT Initiative attempted not only to rearrange space but also the existing power relations between local and national actors and stakeholders in the rainforest. The new discourse, however, enacted the territorial conflict in Yasuní not primarily as a matter of conservation versus use but as a matter of nature conservation versus poverty alleviation and, more specifically, the possibility of reconciling these issues within the same territory. While the Yasuní-ITT Initiative enacted oil extraction in biologically and culturally diverse territories as a highly risky enterprise and was, thereby, aligned with *oil as destruction and violation of rights*, the *developmental mode of ordering* highlights the possibilities of coexistence between extractive industries and nature by enrolling Norway and Canada as successful examples. Consequently, Yasuní-ITT is an illustrative case of just how difficult it can be not only to achieve *interessement* and international *enrollment* (Callon 1986) for this kind of policy shift but also how challenging it can be locally to organize and distribute space between different sectorial interests and conflicting political agendas. The latter is an aspect that the Yasuní and Lofoten cases have in common. The following excerpt from the interview with a representative from Norway's International Climate and Forest Initiative (under the Ministry of Climate and Environment) shows some of the similarities that exist between the ongoing political debates in Norway and in tropical forest countries such as Ecuador:

> The ongoing debate that is taking place in Ecuador, Colombia, Peru, Indonesia, etc. about whether an area should be employed in the production of palm oil, soy, beef or standing rainforest, is not that different from the ongoing debate among the Norwegian central government, municipalities, the county, the oil industry, the fisheries industry, and the local population along the coast. It is the struggle of one interest against another and about optimization of land use in society.
>
> Me: *That is what I also began to realize, that it is not so far from Lofoten to the Amazon.*
>
> No, and it is important for us as Norwegians that we keep in mind that the debate our partner countries have internally is just as difficult for them as the debate about Lofoten–Vesterålen is for Norway. We cannot expect that someone can turn around the land use in Brazil, for example, because Norway is coming in with 2 billion Norwegian crowns a year. It is "pocket money" in relation to what they could actually obtain [from other activities].

Hence, conflicts over resource-rich areas and overlapping territorializations are not that different regardless of whether they are connected to an oil industry that is expanding its operations in Arctic waters or in the tropical rainforest of Amazonia. How these contradictions materially and topographically play out in the two cases is an issue that I believe to be comparable. But before I do so, I want to emphasize that evidently the technological and material conditions for oil operations in Norway and Ecuador are very different, which means that oil is practiced differently. However, the arguments used by national authorities and entities that wish to open new oilfields, even in areas that are highly sensitive due to their complex ecosystems, are not that different. In both cases, they turn to development/welfare to legitimize these lines of action, that is, the countries need oil revenues to finance important public services and welfare for the population. As for today, both countries have highly oil-dependent economies, as neither Ecuador nor Norway has industries or productive sectors that can take over and replace the income obtained from their extractive industries.

While arguments are very similar across sites, they are also partially connected and distributed across levels and, therefore, require a glocal lens to follow their web of connections. The actor-network theory (ANT) approach is useful, because it provides a strategy to overcome the boundedness of predetermined scales with its focus primarily on complex interactions among heterogeneous actors. By moving between sites and scales, "the presumption of cultural or national boundedness that burdens much of the traditional case study research is mitigated" (Bartlett and Vavrus 2017: 77). With this in mind, the vertical axis provides an opportunity to trace how an international policy or event produces negotiation, interaction, adaptation, or resistance at different locations or sites. One such example is when oil prices dropped dramatically in the second half of 2014, which impacted national economies in both Ecuador and Norway, despite the countries' different responses and political strategies to handle the oil crisis. This situation illustrates the importance of studying "the effects of encounters across difference" (Tsing 2005: 6). Although the drop in oil prices was a situation that informants in both countries talked about, the meaning attached to this event varied considerably. Several informants strongly believed that the low oil prices would affect investments in new petroleum projects. In Ecuador, this situation gave hope to informants who were against oil extraction in Yasuní, as they believed the low oil prices would slow down and maybe postpone the development of the ITT block, as the project was not profitable with the current oil prices. The market was believed to work at least as a temporary protection for Yasuní-ITT. Alternatively, in Norway, some stakeholders feared, after further reflection, that depressed oil prices would increase the pressure on LoVeSe, since oil exploitation in the Barents Sea implies dealing with a much more challenging environment that under current conditions, would simply make potential oilfields too expensive to develop and operate. LoVeSe, however, is a less costly project that could probably come on-stream faster. In this case, low oil prices were not perceived as an economic impediment but as a factor that made Lofoten a more viable option, at least from the oil industry's perspective. This illustrates the importance of tracing flows of influence, ideas, and actions across multiple levels (Bartlett and Vavrus 2017: 41) to understand how certain events shape meanings, enactments and potential outcomes on the ground.

The ways in which Lofoten and Yasuní are assembled and inscribed as territories in the Integrated Management Plan for the Barents Sea–Lofoten Area and in the Management Plan for Yasuní National Park present both similarities and differences. I will look into some of the things they have in common first. In both cases, we are dealing with processes of spatial ordering that appear as contradictory, as areas that are inscribed in plan documents and research reports as highly sensitive also simultaneously participate in sectorial territories where various interests and economic activities intersect, such as fisheries, environmental conservation, oil activity, tourism, etc. These competing interests and stakes crisscross Yasuní National Park and the Lofoten plan area and have influenced their processes of spatial ordering. There are zones that not only overlap with each other but also permeate and overflow established boundaries, as the ecosystems do not necessarily comply with their politically designated zones or areas. Species tend to perform their own territorializations according to lifecycles, seasonal variabilities and complex patterns of mobilization. In the case of Lofoten, fish and seabirds enter and exit the plan territory, as there are no ecosystemic border controls but a vast world or worlds in motion. In Yasuní, despite having their own protected areas and the Intangible Zone (Zona Intangible Tagaeri Taromenane; ZITT), humans and nonhumans show up in areas specifically designated and inscribed in maps as oil blocks. These examples of territorial overlaps are not just a local topographical issue, visualized in conflicting temporal and spatial zoning schemes. The territories within also present multiple links and relationships that extend beyond the place itself. In the case of the Barents Sea–Lofoten Area Management Plan, the territorial organization at sea is linked partially to international policies concerned with integrated, ecosystem-based ocean management and partially to a national petroleum sector that strives to access the area. We also find the management plan process distributed in documents and research reports that circulated among several ministries, directorates, research institutes, and civil society organizations. The Steering Committee led by the Ministry of the Environment established an expert group headed by the Norwegian Polar Institute and the Directorate of Fisheries. Together with other entities, representing various sectors[3], the group compiled the existing knowledge base in order to develop the management plan for the Barents Sea–Lofoten area (Ministry of the Environment 2006: 15). Consequently, paying attention across scales means paying attention to networks as they emerge and transform. Yasuní National Park is similarly ordered not only in the Ministry of the Environment that works on conservation plans but also in the Ministry of Justice, which takes care of territorial issues related to the Plan of Precautionary Measures in favor of peoples living in isolation. This document constitutes a partial link to the headquarters of the Inter-American Commission on Human Rights in Washington D.C. We also find many territorializations and re-territorializations in presidential decrees and in reports circulating in the National Assembly. Civil society organizations, together with international environmental networks, also voice their viewpoints regarding the distribution of space through campaigns, projects and local resistance. Within Yasuní National Park, the logic of specific zones and territorializations have historically been rooted in the underground petroleum reservoirs.

Consequently, both cases have to do with territorial conflicts where nature, industries and human stakeholders struggle over space. The ongoing controversies have politically been translated as a question of how to achieve coexistence between productive sectors, cultural identities and various natures. In both cases, the state addresses the issue through spatial ordering as a way to domesticate the conflicting interests and territorialities. In this context, it is important to remember that "spatiality is not given. It is not fixed, a part of the order of things" (Law 2003: 4; emphasis removed), but it is an assemblage that integrates, demarcates and stabilizes the world. The practical work of spatial ordering is always a world-making effort that operates between structures of presence and absence. As Law asserts, Euclidean spatial ordering is just one among many other coexisting possibilities (Law 2003). When I zoomed in on Yasuní's forest and the sea area outside Lofoten, other multiple patterns emerged as other territorialities were simultaneously being performed along with those of the state. These patterns were complex, flowing, disruptive, moving, and often cyclical. These characteristics, however, are not always easy to handle from an administrative or economic point of view. In both cases, the state domesticates the controversies through what can be described as material-semiotic "assembling practices" (Asdal and Hobæk 2016: 99) such as mapping and zoning. Domestication is a syncretic mode of ordering (Law et al. 2013) that operates in a variety of fields. As an ordering device, it is often translated as mechanisms of control and spatial confinement (Lien 2015). Accordingly, domestication's "raw material" is a fuzzy, impure and unfolding world that is ordered through social distribution and temporal and spatial segregation. Along the borders, however, there is always intense activity. All those who are familiar with border towns know that the border, the site between being and being something different, is what brings life to the place. Here you have exchange, traveling, trading and secret smuggling routes. Hence, the proximity to the other place is what makes border towns the perfect location for contact and overflows. It is essentially the demarcation line that produces this booming activity. Border towns can, therefore, work as a useful metaphor for modernity, which "is both pure and it is not pure at all, and that *both/and* is what is distinctive about it and why it is so productive" (Law et al. 2013: 174). As a syncretic mode, domestication is ultimately busy "patrolling the borders" between facts/values; science/politics, and nature/culture.

The inscription of space in both Ecuador and Norway through various mapping practices constitutes an important political tool employed to *tame* the territorial and socioenvironmental conflicts in Lofoten and Yasuní. Maps embody specific perspectives and purposes and are, therefore, not transparent visualizations or neutral knowledge tools. They are political artifacts that impose a particular order upon the territory they are performing and meant to govern; in other words, they are imbued with authority. The Barents Sea–Lofoten area plan territory and Yasuní National Park are both domesticated through the *mode of separation* (Law et al. 2013), that is, by using temporal and spatial distribution. In a rather similar way, the maps seek to produce a unit or a territorial whole by fragmenting the territory in several zones in order to enable coexistence and balance the dispute between various sectors, stakeholders, and industries that are present in these areas. Consequently, there are several logics operating simultaneously in the maps: Economic, environmental, administrative, etc. In both cases, we also find several other logics that are silently being performed. Sometimes, they are partially taken into account, while on other occasions they are deleted or simply ignored. The Northeast Arctic cod participated in organizing the plan territory in Lofoten's case, as the southern boundaries of the plan area were established not only based on administrative considerations but also on the cod's spawning and migration patterns, which connect the Barents Sea to the Lofoten Islands. In specific areas, the fisheries and the oil and gas industry share the same space. But a temporal zone separates them from March 1 to August 31, as exploration drilling is prohibited during this period of the year to protect the fish stocks. Peoples who live in voluntary isolation in Yasuní are confined to a spatial zone, the ZITT, but they continue to perform their vast ancestral territory through their cycles of mobility as they hike through the forest and change their residential locations from time to time. Their ancestral territorialities repeatedly permeate the spatial zone that the state has created to separate their livelihoods and hunting grounds from areas designated to the oil companies. Sometimes there is violence as these indigenous groups defend their territory and perform its borders against the overflow of oil companies, illegal loggers, and other ethnic groups. From this perspective, which is not inscribed in the map, their world is being permeated by surrounding territorial threats, which are both local and translocal. Despite the domestication effort contained in zoning schemes, the place-making strategies of indigenous groups often collide with the spatial ordering performed to accommodate space for national and international oil companies. In Norway, there are also situations where one activity permeates and interferes with other industries or sectors. For example, members of the Norwegian Fishermen's Association often complain that they are displaced from sea areas when oil companies carry out seismic data acquisition. These activities disturb the fish and alter their usual behavior. Overall, the territorial overlaps tend to generate uncertainties regarding long-term effects of seismic surveys on various fish species.

As for zoning strategies, we also find important differences between the cases. The use of temporal zoning is quite different in Yasuní National Park than in the Barents Sea–Lofoten area, where temporal distribution is directly inscribed in the map of the petroleum framework. In the case of Yasuní, there is in fact a temporal zone, but it is situated outside the management plan's field of action. Nevertheless, *criteria of historic use* constitute a temporal relation, which is significant as it enables the further presence of the oil companies in the national park despite article 407 in the Constitution. Overall, this temporal zone largely works as a guarantor for the spatial distribution that currently benefits the oil companies. Here, we see that spatial and temporal rights are closely knit together at the expense of other forms of historic use, such as those of the Waorani and the uncontacted peoples living in Yasuní.

Although there are several important similarities between the two cases as to how conflicts over space are being domesticated through spatial and temporal ordering, and how petroleum resources in the underground become an important drive in this process, there are also interesting differences. In Norway, the political controversy attached to the possibility of opening up the Lofoten area for petroleum activity is also domesticated through scientific research and a continuous process of knowledge gathering. It is highly controversial whether or not oil exploitation should occur in these waters. Science, therefore, becomes a recurrent mechanism by which politics seeks robustness and solidity. Whereas science informs the management plan, and the consensus-based approach was implemented as a way to overcome discrepancies it is, nevertheless, a value-laden process wherein politics and science repeatedly appear as entangled. It is important, therefore, to acknowledge that science and politics participated in a bidirectional dynamic in which scientific knowledge, on the one hand, was translated into political decisions and politics on the other hand, opened up new gaps in the existing knowledge base. In 2006, several uncertainties related to gaps that were identified during the plan process made the government decide not to open up the Lofoten area for petroleum activity during this period. To fill in the missing knowledge, three areas were given priority for the programmed update of the management plan in 2010: Mapping of the seabed, survey and monitoring on the seabird populations, and seismic surveys of petroleum deposits around LoVeSe and the edge of the continental shelf. These areas had been cataloged as particularly valuable and vulnerable, but they were also considered interesting by the oil and gas industry. This interplay between science and politics indicates that while being a knowledge-based process, the Barents Sea–Lofoten Area Management Plan was also strongly guided by politics. Science is never just about facts but is also about values and what is performed at any given time as the common political good.

Scientific advice has a much weaker position in the process related to Yasuní National Park, and it is not used the same way as in Norway, that is, as a tool to domesticate the controversy linked to the possibility of petroleum exploitation in vulnerable and protected areas. In the management plan for Yasuní, the missing link to scientific research is mentioned as one of several identified problems that influence the possibility of proper management of the park. This does not mean that there is little scientific knowledge available. Besides the Galápagos National Park, Yasuní is one of the most researched areas in Ecuador (Ministerio del Ambiente 2011: 18), and two scientific stations belonging to private universities are located within the park area. A constant flow of scientists and students has generated a considerable number of publications and projects during the last two decades, yet there is a lack of coordination with the National Park's administration (Ministerio del Ambiente 2011: 18). According to the Ministry of the Environment, the knowledge that is being produced does not contribute to conservation purposes, the management of the park or the necessities of the local communities (Ministerio del Ambiente 2011: 18). This disarticulation between existing knowledge and Yasuní's management objectives is clearly problematic, as the park's administration operates on a tight budget that complicates the possibility of financing its own research. With the ITT Initiative, however, a kind of re-articulation occurred, as the initiative strived to gain international political robustness by linking Yasuní to existing scientific knowledge from national and international documents, publications and research reports. Whereas the Management Plan for Yasuní National Park emphasizes disconnection from existing research, the Yasuní-ITT Initiative enrolled scientific facts as a way to solidify its own position. The same facts that produced Yasuní as a unique and megadiverse place also allowed the initiative to circulate and extend its web of connections. As I explained in the previous chapter, Yasuní was largely broken down in taxonomies, categorizations and numbers that provided the proposal with increased mobility. Yasuní had to be commensurable to become a unique political good, as its uniqueness or pristine character did not emerge from the place itself but from the multiplying effect of juxtaposition and constant comparison with other similar locations. It all boils down to the position it holds as a biodiversity hotspot when compared by scientists with other similar places in Amazonia's tropical rainforest.

Furthermore, the preponderant role given to science in the Norwegian petroleum debate is clearly occupied by constitutional principles in Ecuador, an important difference between the political processes evolving around LoVeSe and Yasuní-ITT. Whereas science is set to work to settle controversial decisions related to petroleum operations in Norway, the power of the Constitution is invoked in Ecuador. To open up the ITT block for oil drilling, the presidential request to the National Assembly had to achieve a harmonious interaction between the contents of articles 57, 407, and 408 in the Constitution, which was enabled by the new map issued by the Ministry of Justice. In Ecuador, controversial political decisions related to the exploitation of hydrocarbons are not primarily settled in the field of science but in the normative and legal domain. Despite this important difference, the practical consequences for key categorizations are, nevertheless, very similar in the two countries. Several areas are identified in the Integrated Management Plan for the Barents Sea–Lofoten Area, as particularly valuable and vulnerable, where no petroleum activity is permitted at present. These classifications are neither fixed nor completely settled, as they can always be reviewed in the light of updated scientific data. As a rolling document with the possibility of acquiring new attributions, the future management plan could, in theory, modify past categorizations of vulnerable areas within the plan territory:

> The location of the polar front, like that of the marginal ice zone, is being influenced by climate change. In the eastern parts of the Barents Sea, the polar front has receded further north and east. An assessment of whether the delimitation of the polar front as a particularly valuable and vulnerable area needs to be updated will be made when the management plan is revised in 2020.
>
> (Ministry of Climate and Environment 2015: 36)

A possible new delimitation of valuable and vulnerable areas such as the polar front has caused heated debate, especially among environmental non-governmental organizations (NGOs), due not to the scientific data per se, as there are strong indications of the ice retreating but to the implications this new definition could have for the oil and gas industry. They fear that a new delimitation in practice means giving oil companies access into areas that today are classified as sensitive due to their importance for biodiversity and biological production. Several environmental organizations also find expanding oil operations in the Arctic contradictory when combustion of fossil fuels is one of the main causes behind global warming. The environmental status of LoVeSe (Nordland VI, VII, and Troms II) as a particularly valuable and vulnerable area does not constitute a permanent safeguard against future incursions of the petroleum industry. Scientific advice definitely plays a significant role, but so does the political composition in the Norwegian Parliament. It is important, therefore, to take into account that when nature objects are being produced, science does not necessarily move independently or blindfolded, as it is always given some kind of direction as to where and when to scrutinize and when it is supposed to move on.

It is easier to overlook the role played by politics in environmental controversies in Norway due to the country's strong focus on science and knowledge-based decisions. In Ecuador, on the other hand, the process involved in changing the status of protected areas in Yasuní National Park is more clearly political. The fact that the National Assembly must legally substantiate these kinds of decisions brings the normative aspects to the fore, whereas in Norway the interplay between facts and values is much more ambiguous due to the scientific anchorage of political decision-making.

Climate policies as strategic framing

Just like all objects, oil and climate unfold in "creative disjunctions of absence/presence" (Law and Singleton 2005: 348). There must be a series of absences for something to be present. As discussed in the first section of this chapter, oil and climate constitute evolving knowledge objects, due to their unfolding structures of absences. Objects stop being what they are and become something different when these are brought to life. This continuous interplay characterizes the complexity of what Law and Singleton refer to as "messy objects", that is, "objects that cannot be narrated smoothly from a single location" (Law and Singleton 2005: 348). I believe a processual comparative case study is a relevant approach towards this "messiness", as it provides an optic to examine these kinds of objects from various sites and across levels. It gives me the opportunity to present an account of the Norwegian oil and climate dilemmas from Norway as well as from Ecuador, while paying attention to how the international climate framework is negotiated and implemented at the national level. Furthermore, it allows me to describe the interrelatedness of the Amazon crude and global climate change with a view from the rainforest but also as seen from Lofoten and the Arctic. Consequently, I find that multi-sited and multi-scalar projects are particularly well suited to produce accounts about glocal and messy objects.

By following a comparative case study approach towards policy as a set of practices performed by different actors across sites and scales (Barlett and Vavrus 2017), it is possible to zoom in on how different actors define a problem, how they shape solutions and means to resolve it, and how they negotiate visions of the future towards which their policy efforts should be directed (Hamann and Rosen 2011 in Bartlett and Vavrus 2017: 2). Hence, Bartlett and Vavrus emphasize the importance of paying "close attention to how actions at different scales mutually influence one another" (2017: 14). From this perspective comparison becomes a valuable tool to gain insights about the various and often unexpected ways networks and actors are articulated and disarticulated across time and space (Bartlett and Vavrus 2017: 19).

In Ecuador, several informants discussed and compared the rationale behind Reducing Emissions from Deforestation and Forest Degradation (REDD) and the Yasuní-ITT Initiative as climate instruments. They were particularly interested in what they thought of as a more holistic approach towards climate mitigation found in the Ecuadorian proposal as compared to REDD+. Generally speaking, Ecuadorian environmental organizations and large sectors of the indigenous movement tend to be rather skeptical regarding REDD programs, which they claim pose a threat to indigenous territorial rights. This has also been a recurrent critique of *Socio Bosque*, the Ecuadorian forest conservation program.

Furthermore, Norwegian informants representing environmental NGOs and the Ministry of Climate and Environment during interviews also discussed several aspects of REDD+ and its strengths and weaknesses as a climate mitigation mechanism and Norway's key role in its development. They also offered perspectives on why Norway declined contributing to the Yasuní-ITT Trust Fund. This was particularly the case of informants who had been involved in discussions regarding the ITT proposal when representatives of the Ecuadorian government visited Norway. Thus, as data collection proceeded, I realized that various connections and links between Yasuní-ITT and REDD+ were emerging as an important point of reference; in other words, their differences, similarities and overlapping objectives were part of the political debates and controversies attached to how the relationship between petroleum production and climate change was framed and politically handled. On some occasions these links were direct and easily traceable, whereas on other occasions, they had more of an ambiguous or diffuse character. Interestingly, REDD+ came to have a decisive impact on the Yasuní-ITT Initiative's mobilization and enrollment attempts. As a precarious network, the proposal depended heavily on international support and contributions in order to succeed. Consequently, the initiative required relevant actors to support its problem definition and the way it framed the relationship between oil extraction, deforestation and climate change; in other words, it depended on *spokespersons* that could voice its relevance as a new approach towards global climate change mitigation. This aspect was highlighted by a Norwegian informant who argued that the Yasuní Initiative was fragile and, therefore, depended on adherence from significant supporters. Accordingly, it constituted a process in which Norway clearly could have made a difference. Whether Norway's negative response influenced or not the position of other European countries on Yasuní-ITT is difficult to say. Initially, it appeared that Germany would become a significant contributor and advocate for Ecuador's effort to keep the oil in the ground, as its parliament showed both enthusiasm and commitment to support the Yasuní-ITT Trust Fund. After general elections and a new parliamentary composition in the *Bundestag*, however, Germany withdrew from its former commitment in 2010. Dirk Niebel, the new Minister of Economic Cooperation and Development, explained this decision based on a combination of factors: The risk of creating a precedence in the case of other oil-producing countries, the lack of guarantees that the oil would be kept permanently in the ground, and a preference for REDD programs (Martin and Scholz 2014). Both President Correa's insistence on having a Plan B and European donor countries' preference for supporting forest conservation through REDD mechanisms, may probably explain why Germany changed its position (Martin and Scholz 2014). A member of the Administrative Committee for the Yasuní-ITT Initiative reflected on these adverse circumstances:

> He presented [referring to the Ecuadorian president] a discourse as being in favor but in practice the government boycotted the process and, on the other side, I also believe that the countries that could have supported [the initiative] did not have a firm or a clear position. Germany took a very positive stance. There was a change of forces in the parliament and they changed their position drastically, and a government like the Norwegian one that told us from the very beginning that they could not support the project since they exploited oil, and then the reactions of many countries, while being in favor they were not really determined or straightforward.

Seen from Ecuador, Norway's decision not to support the initiative being both an oil producer and a country strongly committed to environmental issues was rather surprising. During one of several visits to Oslo, members from the Administrative Committee for the ITT Initiative were told by government officials that Norway could not make contributions to the Yasuní-ITT Trust Fund the way it was designed but could, alternatively, contribute to other conservation mechanisms. Regarding these matters an official of the initiative stated:

> I was in Oslo and we made a presentation, several presentations to the Norwegian authorities, the Ministry of Cooperation, the Ministry of the Environment as I recall, some other authorities as well. We had presentations with civil society [actors] and we felt very welcome, but in the case of Norway the representatives of the government were very straightforward. I don't know if I should say this but being Norway an oil-producing country and at the same time a country that is very concerned about the environment, they could not support [the initiative] the way the proposal was planned, with a financial contribution to the fund because it would be seen as a contradiction and they were very honest about it. They told us that due to their condition as an oil-producing country a thing like this would be a paradox. Nevertheless, what they said was that they could not contribute to the fund as such, but to projects related to protection, especially in the buffer zone.

While apparently incurring a contradiction, as seen from Ecuador, especially due to the huge support Norway has given to Brazil, the following comment made by a Norwegian petroleum researcher shows that from Norway's perspective, the decision was highly consistent with both national petroleum and climate policies:

> I am convinced that the reason why there was not an agreement concerning Yasuní was the political effect it would have in Norway: Norway financing another country to leave its oil when we ourselves have not left our oil. That is a contradiction, right. Politically, this touched in a way the central problem with the whole climate issue in Norway, namely, being forced to accept that oil production in itself is a problem.

While the Ecuadorian government's political miscalculations and mistakes probably affected the ITT Initiative's capacity to enroll international actors, the explicit focus on oil extraction as an essential part of the climate issue seems to have been an equally important reason for not gaining sufficient support. According to a representative from the Rainforest Foundation (Norway), within ongoing discussions and the international climate framework, it was impossible to receive an initiative like Yasuní-ITT. Compensating a country for not extracting its petroleum resources was an idea that made many actors "uncomfortable", including the Norwegian government. Supporting Yasuní-ITT could be seen as opening up a broader debate regarding which reserves should or "deserve" to be kept in ground; in other words, a debate related to the use of oil-free zones or oil moratoriums as climate tools. In Norway, demands regarding the need to slow down and reduce the petroleum production have increasingly become part of the national climate debate in recent years (Lahn 2017). Environmental NGOs and several political parties claim that there is no "room" for oil from LoVeSe or the Arctic if Norway wishes to fulfill its climate commitment in a 2° C scenario (or the 1.5° C goal from Paris). Overall, the fact that Ecuador and Norway tried to enroll each other due to overlapping interests tied to the rainforest and its role in climate mitigation strategies is interesting. While struggling to translate national and international interests by *reshuffling the means*, the Yasuní-ITT Initiative also evolved as a climate instrument thanks to a continuous and contrasting dialog with REDD+ mechanisms. As REDD programs and Kyoto instruments were actively used to establish similarities and differences with the Yasuní-ITT mechanism by both official documents and informants, I believe it is important to look into some of their connections, framings and overlapping aims and fields of action.

As post-Kyoto proposals, Norway's International Climate and Forest Initiative (NICFI) and the Yasuní-ITT Initiative were launched just a few months apart. Ecuador was first out, as President Correa presented the Yasuní-ITT project for the General Assembly of the United Nations in September 2007, while the Norwegian Prime Minister, Stoltenberg, launched NICFI at COP13 in Bali mid-December the same year. Although the initiatives present several similarities, they clearly frame the climate issue differently, as a situation that demands different approaches towards mitigation. Climate change itself is, thus, produced as a diverse, heterogeneous, and multiple knowledge object. Both the Norwegian and the Ecuadorian initiatives emphasize the problematic relationship between deforestation and global warming. In the proposals, tropical rainforests are performed as important locations where the negative rise in global temperature can be reversed, since tropical forests constitute huge carbon sinks where carbon dioxide is sequestered and stored. According to NICFI (2015), tropical forests "can provide one third of the climate change solution over the next 15 years". The initiatives were aimed at enrolling many of the same countries, due to their common focus on forest conservation as an important factor in combating global warming. As for the ITT proposal, the idea was to start implementing the mechanism in Ecuador as a pilot project, and if it was successful in gaining international acceptance, they would continue replicating the Yasuní model in other developing countries with similar "significant fossil fuel reserves in highly biologically and culturally sensitive areas" (Larrea et al. 2009: 5). The Yasuní-ITT documents specifically mention as possible candidates for this climate mitigation mechanism countries such as Brazil, Colombia, Democratic Republic of Congo, Costa Rica, Indonesia, Peru, Bolivia, Papa New Guinea, etc. Some of these countries are today participating in REDD+ initiatives financed through NICFI. Since 2008, Norway has been one of the key financial contributors to multilateral programs that channel resources to tropical forests countries. A representative from NICFI reflected on the importance of the Norwegian involvement in REDD+ as follows:

> Norway has acquired a position in the international climate debate and in many other forums based on being a big contributor to REDD+. In addition to what the Climate and Forest Initiative has achieved, I dare say that Norway has gained access to actors, access to negotiation processes, and trust from others that we would not have had if we had not led the way with an initiative as big as this commitment is. That is important. It has also been very important in the negotiations leading up to Paris and during [the climate summit in] Paris ... It gives a small country like Norway greater opportunities of influence since we have established trust. We have shown through the Climate and Forest Initiative that we take seriously developing countries' needs for financial support and capacity building, and that gives us a credibility not everybody has.

Besides being a contributor, Norway has also participated actively together with other donor countries to design the multilateral institutional structure around forest initiatives and mechanisms related to global REDD+ processes. This is an important difference between the Norwegian and the Ecuadorian initiatives: As a self-financed mechanism, NICFI had a clear advantage that made it politically easier for Norway to generate *interessement* (Callon 1986) and enroll partnership countries along with multilateral entities than it was for Ecuador, which was requesting contributions from the international community to finance the ITT Initiative. Norway also enjoys a much higher level of trust internationally. The importance of trust-building to enroll potential participants in the Yasuní-ITT network was often stressed among actors and stakeholders in Ecuador:

> The most important currency the government had to ask for money was making them trust us, because you are asking them to trust you "blindly". Trust a project, which means not to extract [the oil] in the future. The way to achieve this was not convincing them how beautiful Yasuní is, but convince them that you are trustworthy, that you are a trustworthy person, or a trustworthy government.

A member of the Yasuní-ITT Administrative Committee stated that emphasis on consistency and on developing an Ecuadorian "green diplomacy" were elements that could have contributed over time to position Ecuador positively in international negotiations. However, I claim that the most important difference between the Norwegian and the Ecuadorian climate mitigation initiatives has to do with the framing process or, in other words, to what extent the national and international frames coincide. In NICFI's case, Norway participated together with other donor countries in designing the "multilateral institutional architecture" (NICFI 2014) related to the UN-REDD/REDD+ framework. Norway was, therefore, largely able to work on and align national and international mechanisms so that NICFI and global REDD+ *fitted smoothly*. Borrowing from Asdal and Moser (2012), this is what I have previously referred to as "contexting". In contrast to contextualizing, which is conceived as placing the issue in a broader surrounding context that is given explanatory power, contexting is about the active work put into shaping and producing simultaneously contexts and issues at stake. This is important, because it constitutes a movement by which particular versions of reality are enacted and, thereby, makes some options or possibilities more plausible or real than others (Asdal and Moser 2012: 303). As with all REDD mechanisms, the Norwegian initiative only considers the link between reduced emissions from deforestation/forest degradation and climate change mitigation. NICFI can be seen as a perfect fit with how the climate issue is framed by UNFCCC and UN-REDD, namely as a carbon emissions problem. Norway has also been successful in expanding the network and enrolling new actors as part of the REDD+ model. In December 2015, during the UN Climate Change Conference in Paris, the Norwegian delegation was given a key role towards the end of the summit to lead some of the difficult negotiations on controversial issues between the parties. One significant achievement from Paris was the inclusion of REDD+ as part of the UNFCCC. The Norwegian Climate and Forest Initiative has actively worked towards this specific goal. Today, as part of the Paris Agreement, REDD+ has become a stabilized mechanism within the international climate regime. Instead of being simply reduced to a component within a broader decision on land use, the REDD mechanism was inscribed and acknowledged with its own article (art. 5.2) in the Paris Agreement. This means that there has been a clear movement from uncertainty and skepticism regarding the possibility of quantifying and reporting on emissions from forests as well as the mechanisms for verifying this information, towards a more general acceptance of the efficiency of the measuring methods that are being used. The NICFI representative emphasized that the cooperation with Brazil and Guyana has demonstrated that it is possible to both quantify and verify these emissions. Although REDD was not formalized as a mechanism until COP13 in Bali (2007), it has roots back to the Kyoto Protocol in art. 2 (a ii):

> Protection and enhancement of sinks and reservoirs of greenhouse gases not controlled by the Montreal Protocol, taking into account its commitments under relevant international environmental agreements; promotion of sustainable forest management practices, afforestation and reforestation.
>
> (United Nations 1998: 2)

The application of these measures was very restricted in the beginning; for example, only afforestation and reforestation activities generated credits under the Clean Development Mechanism (CDM). Another difficulty was the lack of technology to quantify, report, and verify the implementation of such activities (Holloway and Giandomenico 2009). Today, there is increased confidence in the possibility of measuring emissions, which is fundamental to meeting the requirement of result-based payments. This principle has been the mantra of NICFI, and it was also included in the Paris Agreement.

Ecuador's proposal, however, was everything but a perfect fit with the international climate framework and existing forest programs. On the contrary, the Yasuní-ITT Initiative invited a disturbing element into the equation, namely that of fossil fuels, which clearly messed with current accounting practices. As discussed in previous chapters, no climate mitigation mechanism addresses the relationship between extraction of hydrocarbons and climate change. The nation-based accounting scheme starts when fossil fuels have been consumed and combusted, not before they have even been released from the subterranean. The reframing of the climate issue as primarily an extraction problem is a movement that turns several absences into presences by problematizing the role of fossil fuels altogether. In the framing performed by the Ecuadorian initiative, climate change is still linked to the deforestation of tropical rainforests. The focus, however, is on the underground and the vertical territories where fossil fuels are continuously being released and initiate their journey to become future CO_2 emissions. The link to the underground is important, as it broadens and makes explicit the causes of deforestation in these areas. There are problems found above the surface, such as agrarian colonization and illegal logging, but the main drive behind deforestation and forest degradation in the Ecuadorian Amazon is the petroleum resources found in the subsoil. Since the 1970s the oil companies have literally led the way by opening up roads through the forest that follow the pipelines and other petroleum infrastructure. The road axis gives settlers access to previously closed and remote areas, and deforestation continues and multiplies. From a local perspective, current international climate policies seem *unfitting* with the complexity of the rainforest. Hence, the ITT Initiative can be understood as an alternative mechanism that worked on the limitations of the

Kyoto Protocol (Larrea et al. 2009). In a certain way, it is also REDD+ revisited. Instead of seeking financial support for preserving the remaining tropical forests, the Ecuadorian government asked for financial compensation for preserving oil reserves by locking them in indefinitely. Leaving the oil in the ground in the Yasuní-ITT fields would also reduce deforestation and forest degradation, as the initiative would have a multiplying effect.

The process of reframing the relationship between fossil fuels, the rainforest, indigenous peoples, development, and climate change can be interpreted as an attempt on a *counter-enrollment* (Callon and Law 1982) oriented towards providing developing forest countries with an alternative mechanism to participate actively in climate change mitigation (Larrea et al. 2009). The multiple logics that underpin the Yasuní Initiative partially address the local complexities of the rainforest and partially respond to global climate policies. By bringing in non-extraction of fossil fuels as the interface, a component not included in any UNFCCC mechanism, the Yasuní Initiative not only became *unfitting* and incommensurable accounting-wise, but it also caused friction (Tsing 2005). As a hybrid project that worked on several entangled issues simultaneously by employing the current climate toolbox in a rather syncretic way, Yasuní-ITT clearly went beyond the Kyoto Protocol and produced *overflow* (Callon 1998a, 1998b). By enrolling and reframing components from both REDD and the Kyoto Protocol, the initiative struggled to shift the focus from reduction to avoidance of CO_2 emissions. Carbon translations can, therefore, be seen as a strategic means to domesticate complex local encounters between the rainforest, oil reserves and climate change. In other words, it can be seen as a way to make the ITT Initiative a more viable mechanism at the international level.

Over the years, carbon translations have become a key strategy to quantify and manage emissions. As a political, symbolic and moral measure with the capability to bridge places and actions, carbon has come to play a significant role in handling the unboundedness of climate change across scales, from local and specific to global and abstract. Since carbon has increasingly become the dominant optic for thinking about the relationship between humans and the environment (Bridge 2011), the employment of carbon translations became a fundamental mechanism to draw the Yasuní-ITT Initiative together. In this context, carbon translations constituted a specific kind of "infrastructuring" (Blok et al. 2016) of both the petroleum reserves and the rainforest.

By approaching climate change from a completely different angle, that is, going to the source, the Yasuní Initiative engaged in a controversial problematization that is rather unusual for an oil-producing country: The world must start to strand petroleum assets to counter climate change. To demonstrate the value of this thesis the Administrative Committee had to make the initiative accountable. The problem, however, was that the accounting started too early. By attending to *the amounts of avoided CO_2* at the level of the underground, the Yasuní Initiative not only reframed and questioned the reductionism of current climate solutions, but it also worked on national energy politics. As part of a larger political process concerned with the possibility of transitioning towards a less oil-dependent economy, the government and civil society organizations saw the Yasuní-ITT Initiative as a first important step "towards a post-fossil fuel model of development"[4].

Hence, Ecuador's climate initiative was clearly also a matter of national energy policies oriented towards transforming the country's energy matrix. Alternatively, Norway's role as an oil producer is also connected to the country's official climate policies but primarily as structures of absence. Norwegian oil politics are *present as absence*, in other words. However, ordered absences, while necessary to hold an object together or to bound it, so to speak, are all the same potential future presences that new frames can activate and bring to life. With roots back to socioenvironmental conflicts and processes of resistance in Amazonia, the Yasuní-ITT Initiative worked since its beginning to connect non-extraction policies and climate change mitigation. Conversely, Norwegian policies seem to be working in the opposite direction, that is, performing a decoupling between the question of further expansion of petroleum operations and GHG emissions (Andersen 2016: 270). This is an important difference between the Ecuadorian Yasuní-ITT and the Norwegian REDD+/NICFI proposals. Due to the use of different frames, different relationships and connections are taken into account that actually produce not only different climate issues but also different ways of addressing these issues. Norway's "doublethink" has been possible thanks to the international quota system and carbon trading. With an oil-dependent economy and few possibilities of performing inexpensive emission cuts at home due to the extensive use of hydropower, Norway strongly favored carbon trading across borders ever since the international climate negotiations initiated in the early 1990s. (Asdal 2011; Hermansen and Kasa 2014).

The Kyoto regime's consumption-based accounting has enabled Norway to expand its oil operations, reach its emissions reduction commitment, and at the same time have a strong presence in international climate negotiations. As a net exporter, Norwegian oil is mostly consumed elsewhere and does not enter the national emissions accounting. However, being a major oil producer can appear as contradictory when combined with the ambition of a leading role in climate matters (Hermansen 2015). Against this background, the strong political support towards NICFI has been explained as a combination of several factors: First, it harmonized with the government's objective of cost-efficient climate solutions, as tangible results could be obtained relatively fast. Second, a strong initial involvement of two important environmental NGOs, the Rainforest Foundation (Norway) and Friends of the Earth (Norway), gave legitimacy to the process, besides channeling an increasing demand for concrete action regarding climate policies in a context where Norway was lacking cheap domestic options (Hermansen and Kasa 2014).

I believe there is another important side to the general acceptance of NICFI among political parties in Norway, namely, the possibility of taking on an active role in international climate mitigation without having to stumble over complicating and controversial issues. Accordingly, it has been suggested that Norway's enthusiasm for carbon sequestration mechanisms (REDD+ and CCS) seems to be rooted in specific national needs due to "the combination of irrefutable petroleum interests and normative ambitions" (Røttereng 2018: 217) at the national and international levels. Overall, REDD+ processes are compatible with the way Norway frames climate change, namely, as a question of negative externalities. The economic logic that underpins this perspective translates the problem as a situation that arises after the goods have been produced and consumed, which means that there is nothing wrong with the goods themselves. Moving into the petroleum sector, the problem is not found in oil production but appears as an unfortunate side effect when oil is being combusted at the end of the chain (Princen et al. 2015). In short, Norway produces the goods, the "bads" are produced somewhere else.

While NICFI/REDD+ is not linked to the LoVeSe controversy, it allows Norway to participate in global climate change mitigation without politically compromising or foreclosing the opening of certain oilfields in the future. As there is no direct link between opening areas such as Lofoten and paying for the conservation of tropical forests, REDD+ initiatives are enabling because they hold possibilities politically open. We are back to domestication, in this case by sectorial distribution, as rainforests and petroleum belong to separate ministries and frameworks. From this perspective, signing up for the Yasuní-ITT Initiative would have been politically risky due to the performativity of framings. In this case it would imply acknowledging that the climate issue is deeply rooted in fossil-fuel reservoirs, a situation that requires much more drastic measures than just managing emissions at the end of pipe (Princen et al. 2015). Overall, paying to keep the oil in the ground as a means to preserve biodiversity and combat global climate change would make the opening of the Lofoten area (or any other sensitive area with high biological value) almost impossible without risking a situation of double discourse and political inconsistencies.

Ecologization and/or modernization

In both cases, tension arises from the copresence of several natures that operate and are produced in interaction with legal frameworks, economy, environmental policies, knowledge systems, and diverse territorial practices. In order to approach these natures in a more systematic manner I will draw on Latour's (2007) pairing concepts *modernization* and *ecologization*, conceiving them as two ideal types to illustrate specific differences as to how nature-human relationships are experienced and politically handled. When using the Weberian concept of ideal types, I refer to how certain practices can be associated more specifically with modernization efforts, that is, acting towards the world as split into a human and a nonhuman domain, while other policies lean towards ecologization, wherein complex nature-culture imbroglios are taken into account and acted upon in practice. However, when analyzing the controversies evolving around Lofoten and Yasuní, I realized that often it can be difficult to classify network processes as either/or, since ecologization and modernization often occur side by side, creating a series of conflicting policy outcomes.

Since the discovery of oil in the Ecuadorian Amazon, state policies have displayed a modernization drive oriented towards national integration of what was to become an important economic territory. Building roads and granting land to settlers became key strategies used by the state to integrate the region with the rest of the country. As part of this process, various governments conceived the presence of the oil companies as an important mechanism to achieve national development and modernization. With time, the oil companies also became strong actors, as they largely came to replace the state in many of its functions towards the local inhabitants. Overall, oil operations in Amazonia are associated with a process of domesticating a "savage" and "inhospitable" nature as well as its peoples. The forest itself was, to a large extent, an obstacle that had to be cut down as it literally stood in the way of economic activity by limiting access, transportation and circulation of oil and other commodities. In this case, domestication constitutes "a powerful image of how humans have transformed their nonhuman surroundings" (Lien 2015: 7). *Capitalist nature* (Escobar 1999) was being produced through new legal frameworks, such as the law for colonization of the Ecuadorian Amazon (1978), oil concessions, roads and petroleum infrastructure. This process is linked to what I have previously denominated *the developmental mode of ordering*, where nature is taken into consideration primarily as natural resources that generate important income for the state and the possibilities of financing development. In the Amazon region, modernization is specifically displayed in the way capitalist nature is ordered and assembled as productive space, often at the expense of other natures and place-making strategies.

In the northern Amazon, severe degradation of the rainforest, and its impact on the livelihoods of people living in the vicinity of petroleum installations, has over the years, generated resistance among indigenous communities, *colonos* and environmental NGOs. This situation has increasingly produced new ways of ordering the Amazon crude, objecting to the aggressive exploitation of natural resources and the lack of environmental standards in industrial processes where nature is being transformed and commodified. Here, the ongoing lawsuit against Texaco-Chevron is a powerful narrative that links socioenvironmental risk to the lived experiences of the local population. In the case of Yasuní-ITT, environmental risk was always conceptualized by looking to past experiences with the toxic materiality of oil and the use of obsolete technology. In this case, there were no technical heroes.

Political processes at the national level, strongly motivated by civil society demands, produced a broad public debate regarding the role of extractive industries in developmental policies. Here we find a strong link to the modes of ordering that I have referred to *oil as destruction* and *oil as violation of rights*. These modes emerged as alternative ways of ordering and reflecting on how humans and nature are entangled in multiple and often dangerous and uncertain relationships. In Ecuador, we find the most eloquent attempt of ecologization in the constitutional process leading up to the new Constitution in 2008. The inscription of Nature/ Earth as a legal subject with constitutional rights clearly generated tension with other natures that are present and at work in the supreme law. My point being that while *Pachamama*, a living organic being (Estermann 1998) experienced by humans through ceremonies, reciprocity and cultivation of the land, is given considerable space in the legal text, this *organic nature* (Escobar 1999) coexists with other more instrumental and economic versions. This situation generates tension and a series of legal contradictions. While the *Rights of Nature* present various limitations when it comes to implementation, they are important, nevertheless, as an ethical benchmark against which socioenvironmental conflicts in Ecuador are increasingly being discussed and viewed. Some of the apparent contradictions present in the Yasuní-ITT Initiative responded to an intricate combination of both modernization and ecologization, which can be seen in the way the initiative enacted both a pristine and untouched nature and a nature entangled in oil exploitation, deforestation and human settlement. Consequently, the Yasuní-ITT Initiative produced several natures simultaneously, which makes it appropriate to talk about a politics of hybrid natures (Escobar 1999), which is "a way of crossing the boundary between the traditional and the modern" by "using both local and transnational cultural resources" (Escobar 1999: 13).

Moreover, the entangled biocentric and anthropocentric natures found in the Ecuadorian Constitution are truly political and ecological in a Latourian sense, as they show some of the challenges related to defining more symmetrical and respectful relations between humans and nonhumans. In the Constitution's preamble, the construction of a new form of public coexistence is linked to the idea of *Sumak Kawsay*, "the good way of living", which can only be achieved in diversity and harmony with nature. To a certain extent, the Constituent Assembly resembled Latour's (1993) *Parliament of Things* while working on the institutionalization of nature's rights. In a network of statements, politicians, debates, NGOs, the indigenous movement, hearings, lawyers, environmental reports and texts, national media, etc., *Pachamama* gained adherence and expanded, as several organizations and actors used the Assembly to voice their opinion and expose the importance of giving nature subject status as a strategy to address the current ecological crisis. Later, the *Rights of Nature* became one of the main arguments enrolled by the Yasunidos collective in its attempt to pull through a national referendum on oil extraction in Yasuní-ITT.

In Norway, the movement between modernization and ecologization does not reach the same level of conflict as in Ecuador, where tension repeatedly surfaces due to the legalization of nature as a subject of law and the controversy over the political implications of this status. Overall, Norway's petroleum history is marked by a strong faith in the possibility of mastering offshore petroleum operations in challenging environments through political decision-making and regulatory frameworks. When oil was first discovered on the Norwegian Continental Shelf, emphasis was placed on economic measures as a means to refrain from heating up the economy, as well as on the development of technological competence and know-how. From the state's perspective, this was fundamental to building up a national petroleum sector and mastering the various sides of the industry. With the new oil wealth also came the possibilities of progress and economic resources that enabled extending the services of the welfare state and, thereby, transforming the Norwegian society. *Oil as safety* is, therefore, an ordering mode concerned with economic growth and welfare. Hence, progress and stability can be achieved as long as the petroleum resources are handled correctly through proper planning, economic calculation and technological expertise. In Norway's case, accounts about risk also look to the past but, contrary to what happens in Ecuador, there is a lot of storytelling about "technical heroism". Furthermore, environmental risks are often translated as technological possibilities and, therefore, not necessarily a bad thing, as this motivates innovation and creative solutions.

Besides a different environment, stronger safety measures and the use of less invasive technology, the nature that is being spatially ordered, distributed and technologically domesticated is the same *capitalist nature* that we encountered in Ecuador. This passive and objectified nature was born in post-Renaissance Europe and further developed in late 18th-century capitalism (Escobar 1999: 6). As a dominant regime that reaches most parts of the world, it can be difficult to catch a glimpse of other natures, especially in technologically and scientifically advanced western societies such as Norway. Although we do not find an organic Mother Earth in the Norwegian petroleum debate, we do find a series of hybrid natures that are instrumental but also organic, locally anchored and displayed in identity and cultural expressions. For people along the coast of northern Norway, the surrounding environment and the fisheries are not just a basis of economic activity; they have adapted and organized around nature's patterns and seasonal cycles for generations. In Lofoten, the cod's yearly migration from the Barents Sea is the reason why human settlements exist in the area. People feel they belong to a place-based nature, which has given them and their ancestors not only livelihood opportunities but also a way of life. This way of living and interacting with nature often collides with *pure capitalist nature* such as the one performed by the oil industry, which is normally emptied of local and cultural connectivity.

If we zoom in on national policies and documents, looking for ecologization efforts, we will find some examples. While not abandoning the sweeping force of modernization, there are several matters of concern that have generated discontinuity and rupture with previous ways of understanding the relationship between oil and the environment. In the early days of the Norwegian oil industry, moving beyond latitude 62 was considered primarily a technical challenge. Politicians regarded crossing this boundary as a practical question. With white paper 25 (1974) issued by the Ministry of Finance, possible environmental risks related to offshore petroleum exploitation were directly taken into account, besides focusing on the possibility of territorial conflicts between the fisheries and the oil industry. Consequently, latitude 62 went from being a technical frontier to becoming a much more complex environmental and political issue. The transition from oil as a question of mastering through technology and knowledge towards becoming a more uncertain reality linked to environmental conditions marks the transition of *oil as safety* towards *oil as insecurity*. The latter is about economic uncertainties due to depressed oil prices, where the market has made oil a less stable and robust object, but it is also entangled in global climate change. Consequently, *oil as insecurity* is preoccupied with the way climate change is rearranging oil as a less coherent and solid object than it used to be. There is tension between *oil as safety* and *oil as insecurity*, as insecurity inhabits safety as a latent possibility. As an ordered absence ready to take over as presence, *oil as insecurity* is displaying the role of *safety* in delaying the *green shift*. Overall, there is more "ecologization potential" in insecurity as an ordering mode than in the case of safety.

Kodak moments compared

In Norway, many informants believe that the *green shift* is still an open process that lacks orientation and direct involvement on behalf of the state. There are policies that locate the transition towards a low-carbon economy in the transportation sector, housing, industry, land use, and technology such as CCS. Emphasis is also placed on markets and their role in making "green" technology more competitive and available for consumers. Regarding the oil and gas industry that is responsible for a large part of the Norwegian CO_2 emissions, there are few if any specific measures to start phasing out the industry, a necessary step if Norway is going to reach the politically established goal of being carbon-neutral by 2050. Many informants described the situation as a Kodak moment when Norway, instead of seizing the opportunity of becoming less oil dependent and diversifying the economy, continues expanding the oil frontier in the north. The last two concessions rounds are strong indications of an industry that is securing its existence for several decades to come. I have called this process *out-framing*, which differs from the concept of externalities which refers to all those relations not accounted for. On the contrary, in the case of *out-frames*, these relations are highly considered by locating them as part of another context. In Norway, this situation translates into policies that decouple oil and climate as two separate issues. By opening up new oilfields and locking in labor, investments, research, and technological innovation, Norway could, thus, lose the opportunity to take on a leading role in the development of renewable energy and alternative technological solutions.

Whereas Norway's Kodak moment is about policies that keep oil and climate apart, despite scientific knowledge that link them together, Ecuador's Kodak moment is the outcome of policies that worked in the opposite direction, namely linking oil extraction to climate change. As a public policy, the Yasuní-ITT Initiative was conceived as an important step towards less reliance on petroleum resources and a way to initiate a transition towards a post-petroleum development model. This change of paradigm would require a transformation of the country's energy matrix, and the focus was, therefore, on replacing carbon-intensive power generation with renewable energy sources in combination with technological innovation and social programs. When the government decided to terminate the initiative and move towards the implementation of plan B, this situation was regarded by many informants as a lost opportunity to change the course in national petroleum and environmental policies. According to several informants, the Yasuní-ITT Initiative constituted a unique moment to pursue a more sustainable model of development less dependent on extractive industries. Ecuador had reached a political crossroads, in other words, and Yasuní was a long-awaited chance to turn the country's energy policies in a new direction. Rephrasing Latour (2007), the decision was whether to modernize or to "yasunize"[5]. Many civil society actors argued that instead of implementing plan B, Ecuador should have continued with the ITT Initiative even without financial contributions to the trust fund. In the end, they argued, the country would have gained much more from making this kind of statement to the international community than from opening up this pristine environment to oil drilling.

Notes

1   El Niño and La Niña, also known as El Niño-Southern Oscillation (ENSO) cycle, are fluctuations in temperature (warm and cold phases) between the ocean and the atmosphere in the Equatorial Pacific. El Niño and La Niña constitute deviations from normal surface temperatures, which can influence global weather conditions and climate. National Ocean Service (Updated 2021, February 26): http://oceanservice.noaa.gov/facts/ninonina.html

2   Translation by the author.
    Speech of President Rafael Correa (August 15, 2013): www.presidencia.gob.ec

3   The other entities that were integrated into the expert group were the Institute of Marine Research, the Norwegian Petroleum Directorate, the Norwegian Coastal Administration, the Norwegian Pollution Control Authority, the Directorate for Nature Management, the Norwegian Maritime Directorate and the Norwegian Radiation Protection Authority (Ministry of the Environment 2006: 15).

4   UNDP Multi-Partner Trust Fund Office: Yasuní-ITT Fact Sheet: http://mptf.undp.org/yasuni

5   The verb «yasunizar» was invented by environmental organizations to emphasize what was at stake with the ITT Initiative. According to Martínez Alier (2013), Acción Ecológica wrote to the Royal Spanish Academy to get the word accepted in the official Spanish dictionary.

References

Andersen, G. (2016). *Parlamentets Natur: Produksjon av Legitim Miljø og Petroleumspolitikk (1945–2013)* [Doctoral thesis, University of Bergen, Bergen].

Asdal, K. (2011). *Politikkens Natur-Naturens Politikk.* Oslo: Universitetsforlaget.

Asdal, K., & Hobæk, B. (2016). Assembling the whale: Parliaments in the politics of nature. *Science as Culture*, 25(1), 96–116.

Asdal, K., & Moser, I. (2012). Experiments in context and contexting. *Science, Technology, & Human Values*, 37(4), 291–306.

Bartlett, L., & Vavrus, F. (2017). *Rethinking Case Study Research: A Comparative Approach.* London: Routledge.

Blok, A., Nakazora, M., & Winthereik, B. R. (2016). Infrastructuring environments. *Science as Culture*, 25(1), 1–22.

Bradshaw, M. (2014). *Global Energy Dilemmas.* Cambridge: Polity Press.

Bridge, G. (2011). Resource geographies I: Making carbon economies, old and new. *Progress in Human Geography*, 35(6), 820–834.

Callon, M. (1986). Some elements of a sociology of translation: Domestication of the scallops and the fishermen of St. Brieuc Bay. In: J. Law (ed), *Power, Action and Belief: A New Sociology of Knowledge?* (pp. 196–233). London: Routledge & Kegan Paul.

Callon, M. (1998a). An essay on framing and overflowing: Economic externalities revisited by sociology. In: M. Callon (ed), *The Laws of the Markets* (pp. 244–269). Oxford: Blackwell Publishers.

Callon, M. (1998b). Introduction: The embeddedness of economic markets in economics. In: M. Callon (ed), *The Laws of the Markets* (pp. 1–57). Oxford: Blackwell Publishers.

Callon, M., & Law, J. (1982). On interests and their transformation: Enrolment and counter-enrolment. *Social Studies of Science*, 12(4), 615–625.

Creswell, J. W. (2013). *Qualitative Inquiry & Research Design: Choosing among Five Approaches*. Los Angeles. Sage.

E24. (2017, March 31). *Ber de store partiene verne om oljebransjen*. Retrieved from http://e24. no/energi/norsk-olje-og-gass/ber-de-store-partiene-verne-oljebransjen/23962956.

El Diario.ec. (2007, November 5). *Ministro noruego viene a hablar de petróleo*. Retrieved from https://www.eldiario.ec/noticias-manabi-ecuador/58663-ministro-noruego-viene-a-h ablar-de-petroleo/.

Escobar, A. (1999). After nature: Steps to an antiessentialist political ecology. *Current Anthropology*, 40(1), 1–30.

Escobar, A. (2001). Culture sits in places: Reflections on globalism and subaltern strategies of localization. *Political Geography*, 20(2), 139–174.

Escobar, A. (2011). Sustainability: Design for the pluriverse. *Development*, 54(2), 137–140.

Estermann, J. (1998). *Filosofía Andina: Estudio Intercultural de la Sabiduría Autóctona Andina*. Quito: Abya-Yala.

Hermansen, E. A. (2015). I will write a letter and change the world: The knowledge base kick-starting Norway's rainforest initiative. *Nordic Journal of Science & Technology Studies*, 3(2), 34–46.

Hermansen, E. A., & Kasa, S. (2014). Climate policy constraints and NGO entrepreneurship: The story of Norway's leadership in REDD + financing, Working Paper 389. *CGD Climate and Forest Paper Series*, 15 (pp. 1–31).

Holloway, V., & Giandomenico, E. (2009, December 4). *Carbon Planet White Paper: The History of REDD Policy*. Retrieved from https://redd.unfccc.int/uploads/2_164_redd _20091216_carbon_planet_the_history_of_redd_carbon_planet.pdf.

Klassekampen (2007, October 18). *Utfordrer Solheim*. Retrieved from http://www.klassekam pen.no/48229/article/item/null/utfordrer-solheim.

Knorr Cetina, K. (2001). Objectual practice. In: T. R. Schatzki, K. Knorr Cetina, & E. Von Savigny (eds), *The Practice Turn in Contemporary Theory* (pp. 175–188). New York: Routledge.

Lahn, B. (2017). *Redusert oljeutvinning som klimatiltak: Faglige og politiske perspektiver*. Oslo: CICERO Senter for Klimaforskning.

Larrea, C., Greene, N., Rival, L., Sevilla, E., & Warnars, L. (2009). *Yasuní-ITT Initiative: A big idea from a small country*. Retrieved from http://mptf.undp.org/document/download /4545.

Latour, B. (1993). *We Have Never Been Modern*. Cambridge, MA: Harvard University Press.

Latour, B. (2007). To modernize or to ecologize? That is the question. In: K. Asdal, B. Brenna, & I. Moser (eds), *Technoscience: The Politics of Interventions* (pp. 249–272). Oslo: Unipub.

Law, J. (2003). *Topology and the Naming of Complexity*. Retrieved from Centre of Science Studies. Lancaster University: http://www.lancaster.ac.uk/fass/resources/sociology-online-papers/papers/law-topology-and-complexity.pdf.

Law, J. (2007). *Actor Network Theory and Material Semiotics*. Retrieved from http://www.hete rogeneities.net/publications/Law2007ANTandMaterialSemiotics.pdf.

Law, J., & Singleton, V. (2005). Object lessons. *Organization*, 12(3), 331–355.

Law, J., Afdal, G., Asdal, K., Lin, W.-y., Moser, I., & Singleton, V. (2013). Modes of syncretism: Notes on noncoherence. *Common Knowledge*, 20(1), 172–192.

Lien, M. E. (2015). *Becoming Salmon: Aquaculture and the Domestication of a Fish*. Oakland, CA: University of California Press.

Martin, P. L., & Scholz, I. (2014). *Policy Debate. Ecuador's Yasuní-ITT Initiative: What Can We Learn from Its Failure?* Retrieved from https://journals.openedition.org/poldev/1705.

Martínez Alier, J. (2013). Venturas y desventuras de la Iniciativa Yasuní-ITT. In: J. Martínez Alier, E. Gudynas, J. Parrilla, I. Carvajal, A. Rosero, & R. Báez (eds), *Sacralización y Desacralización del Yasuní* (pp. 11–30). Quito: Centro de Pensamiento Crítico.

Ministerio del Ambiente (2011). *Plan de Manejo del Parque Nacional Yasuní*. Quito, Ecuador Retrieved from http://suia.ambiente.gob.ec/documents/10179/242256/45+PLAN +DE+MANEJO+YASUNI.pdf/8da03f55-1880-4704-800e-d5167c80089c.

Ministry of Climate and Environment (2015, April 24). *Update of the Integrated Management Plan for the Barents Sea-Lofoten Area Including an Update of the Delimitation of the Marginal Ice Zone. Meld. St. (2014–2015) Report to the Storting (White Paper)*. Retrieved from https://ww w.regjeringen.no/contentassets/d6743df219c74ea198e50d9778720e5a/en-gb/pdfs/ stm201420150020000engpdfs.pdf.

Ministry of the Environment (2006, March 31). *Report No. 8 to the Storting. Integrated Management of the Marine Environment of the Barents Sea and the Sea Areas off the Lofoten Islands*. Retrieved from https://www.regjeringen.no/globalassets/upload/md/vedlegg/stm200520060 008en_pdf.pdf.

National Ocean Service – National Oceanic and Atmospheric Administration, U.S. Department of Commerce (Updated 2021, February 26). *What are El Niño and La Niña?* Retrieved from http://oceanservice.noaa.gov/facts/ninonina.html.

NICFI – Ministry of Climate and Environment (2014, December 14). *Multilateral Collaboration*. Retrieved from https://www.regjeringen.no/en/topics/climate-and-e nvironment/climate/climate-and-forest-initiative/kos-innsikt/fn-og-verdensbanken/ id2344812/.

NICFI – Ministry of Climate and Environment (updated 2015, February 20). *Why NICFI and REDD+?* Retrieved from https://www.regjeringen.no/en/topics/climate-and-e nvironment/climate/climate-and-forest-initiative/kos-innsikt/hvorfor-norsk-regns kogsatsing/id2076569/.

Norwegian Agency for Development Cooperation (2011, October 29). *How We Work*. Retrieved from https://www.norad.no/en/front/thematic-areas/oil-for-development/ how-we-work/.

NORAD (2013, January). *Facing the Resource Curse: Norway's Oil for Development Program*. Retrieved from https://www.norad.no/globalassets/import-2162015-80434-am/www. norad.no-ny/filarkiv/evalueringsavdelingens-filer/facing-the-resource-curse_norways- oil-for-development-program.pdf.

Presidencia de la República. (2013, August 15). *Anuncio a la Nación Iniciativa Yasuní ITT*. Retrieved from http://www.presidencia.gob.ec/wp-content/uploads/downloads/2 013/08/2013-08-15-AnuncioYasuni.pdf.

Princen, T., Manno, J. P., & Martin, P. L. (2015). The problem. In: T. Princen, J. P. Manno, & P. L. Martin (eds), *Ending the Fossil Fuel Era* (pp. 3–36). Cambridge, MA: MIT Press.

Røttereng, J.-K. S. (2018). When climate policy meets foreign policy: Pioneering and national interest in Norway's mitigation strategy. *Energy Research & Social Science*, 39, 216–225.

Tsing, A. L. (2005). *Friction: An Ethnography of Global Connection*. Princeton, NJ: Princeton University Press.

UNDP Multi-Partner Trust Fund Office (n.d.). *Yasuní-ITT Fact Sheet*. Retrieved from http://mptf.undp.org/yasuni.

United Nations (1998). *Kyoto Protocol to the United Nations Framework Convention on Climate Change*. Retrieved from https://unfccc.int/resource/docs/convkp/kpeng.pdf.

World Commission on Environment and Development (1987). *Our Common Future*. Oxford: Oxford University Press.

7 Enactments between presence and absence

A conclusion

The research project on which this book is based was organized around the analysis and comparison of two socioenvironmental controversies related to possible petroleum extraction in areas that are cataloged by science as highly sensitive. The cases, Lofoten and Yasuní-ITT (Ishpingo, Tambococha, and Tiputini), were conceived as political processes and not as bounded geographic places. However, the characteristics and attributes of these locations inform and orient the petroleum debates in both Norway and Ecuador and beyond their borders. The cases are therefore place-based but not place-bound and can be described as ongoing processes that are assembled simultaneously at different sites. Their continuous movements, displacements, assembling, and reassembling across scales called for an emergent approach. Thus, the boundaries of the cases were not defined a priori but empirically investigated. Accordingly, *what* and *who* made up the cases was a question of constant revision and analysis as the research proceeded. Despite my interest in comparing cases from a developed and a developing country with highly oil-dependent economies, placed at opposite sides of the spectrum in the resource curse literature, the cases were not only theoretically defined but also defined in terms of the comparative case study methodology I chose to work with. Emphasis was placed, therefore, on tracing connections, actors, policies, and practices as the inquiry unfolded (Bartlett and Vavrus 2017). Furthermore, the selection of an actor-network theory (ANT) approach is also largely responsible for what the cases finally became, as this framework refuses to explain phenomena and outcomes in terms of categories and concepts that are commonly used in the social sciences. Conversely, agency, structure, the micro, the macro, nature, society, subject, object, etc., are not treated as fixed ontological realities but as realities that are produced and come into being through processes and transforming networks.

Here, I will revisit the empirical chapters to review the main findings and discuss some of my analytic choices. I then discuss the possible contributions and limitations of the inquiry as well as offer some reflections regarding my own role and positioning as a researcher. Finally, I conclude by reflecting on the analytic potential of the processual comparative case study methodology.

Revisiting empirical findings and analytic choices

The purpose of this book has been to examine how oil and climate change are complexly linked and entangled and how this interrelatedness is handled through knowledge and policy practices at different levels. The main question that guided the research was the following: How is the relationship between petroleum production and climate change politically framed and negotiated in contested cases?

The analysis indicated that this relationship is neither fixed nor stable but rather moves between *presences* and *absences* (Knorr Cetina 2001; Law 2004), enacted by different *framings* (Callon 1998a, 1998b). In both cases, actors and stakeholders acknowledge that fossil-fuel consumption plays a fundamental role in global warming. As such, the authority of scientific knowledge is not disputed or challenged. Conversely, what produces debate are different ways of framing the relationship between oil and climate change by different actors, policies, documents, and sectorial interests. When analyzing and comparing empirical data from the Lofoten and Yasuní-ITT cases, I have discussed how climate change becomes different issues in different policy practices. This *plasticity* is not primarily rooted in scientific discrepancies or ambiguities but rather shows that there is no such thing as universal knowledge, as knowledge has to travel and circulate to be acted upon and implemented locally, which necessarily implies "modifying work" (Asdal 2015). According to Strathern, anthropologists have traditionally been "alert to the nontranslatability of different types of knowledge across conceptual universes" (Diemberger et al. 2012: 239). A central argument of this book, however, is not the nontranslatability between worlds but rather how climate knowledge specifically moves and circulates between these worlds through translations. These movements always imply transformation and *betrayal* (Law 2003b). Consequently, it is not just a matter of different perspectives, that is, seeing the world from different viewpoints, but a question of how climate change multiplies and becomes different political issues. As part of this analysis, I discussed how translations produce displacements and sometimes even *overflow* (Callon 1998a, 1998b). Thus, in one way or another, knowledge is never all-encompassing or universal but situated (Haraway 1988). As this inquiry demonstrates, climate science and the international climate framework are negotiated and framed differently at various levels and sites. As a consequence, international climate policies are not simply transferred; they are continuously negotiated, adapted, resisted, reassembled, and coordinated with other national policies and political aims. This situation produces not only different climate issues but also different climate solutions.

In Chapter 2, the analysis suggested that oil cannot be understood as a single or coherent "thing" but as a complex reality that multiplies in different political, technological, environmental, and economic practices. Contrary to what I expected to find, the informants' accounts about the Norwegian petroleum industry and its future challenges did not primarily vary because of the informants' roles or degrees of environmental commitment, but rather due to different contexts that were evoked and woven into the oil issue. When analyzing the data, two dominant patterns emerged: *Oil as safety* and *oil as insecurity*. To make sense of these patterns, I turned to Law's (1994) modes of ordering. While the concept was originally used "to explore the ordering implied in organization" (Law 2003a: 1), I realized that it could be fruitfully applied when analyzing the various ways in which oil was arranged and coordinated in both documents and interview data. According to Law, modes of ordering are heterogeneous arrangements and processes understood as strategies, which tend to be multiple rather than single (Law 2003a: 1–2). Between different ordering logics, there are often complex relations that depend on specific material and discursive arrangements to operate and perform (Law and Moser 2003; Law 2003a).

When oil in the Norwegian case was ordered as *safety*, it was linked to the welfare state and economic growth and stability. This ordering mode presents oil as a powerful agent that participates in several sectors and areas of Norwegian society by modifying *the state of affairs* (Latour 2005). Through firm planning and a well-designed petroleum framework, the oil revenues have been distributed across the country and, thereby, benefited the population at large. Norwegian petroleum policies have been inclusive, in other words. This success is largely explained by looking to the past. By enacting oil as a turning point in the country's economic history, this mode of ordering coordinates several factors that participated in performing *the state of mastering* the oil industry, such as expert knowledge, policy practices, and technological innovation. The analysis of key documents, such as white paper 25 (1973–1974), demonstrated how uncertainties regarding the incipient petroleum industry in the early years were worked upon and stabilized through political and legal frameworks, economic calculations, and technical know-how. From the beginning, oil was turned into a knowledge object, often at the expense of other areas and research fields. In this ordering mode, there is a lot of storytelling about economic and technological success.

Oil as *insecurity*, however, is an ordering mode that emphasizes how things have changed within the petroleum sector, which has made oil increasingly uncertain and unstable. Informants highlighted circumstances such as depressed oil prices and reduced investments in the oil and gas sector. The chapter found that climate change is adding to this uncertainty, as climate science is making national petroleum policies increasingly problematic and fuzzy. How different scales operate in a relational dynamic in which one level can destabilize what happens at another one is also part of this ordering mode. This illustrates the importance of working across scales, as uncertainties in the petroleum debate are often a scalar matter.

While *oil as safety* and *oil as insecurity* seem to work as opposite modes, the analysis showed that this was not quite the case. *Safety* is deeply involved in the enactment of *insecurity*. What is being coordinated within this mode, besides low oil prices and uncertainties related to global climate change, is the material and discursive content of *oil as safety*. This was illustrated by informants who argued that it was oil's success and former stability that was producing the ambivalence that was delaying the green shift. These modes are, therefore, interrelated since safety participates in how insecurity frames and coordinates oil. Safety becomes key ordering material for insecurity, in other words, which points to a regular pattern of deletion (Law 1994), as the oil and gas industry is strangely absent in official policies and storylines about the green shift. Thus, the analysis demonstrated the various ways insecurity inhabits safety as a latent possibility that can be activated and brought to life at any time. This indicates that ordering should be viewed as an intersecting performance in which multiple discourses or logics intervene (Law and Moser 2003: 7).

Chapter 2 also discussed the relevance of framing (Callon 1998a, 1998b) as an analytic approach towards national and international climate negotiations and policies. As a performative practice, different framings produce different climate issues and, thereby, also different ways to address mitigation (illustrated by the Yasuní-ITT case). From this perspective, framing is what specifically produces an issue as an issue, disentangling it from certain types of knowledge, relations, actors, effects, etc., which are situated outside the frame and, therefore, not accounted for (Callon 1998a). As part of the analysis, the chapter examined the way Norway uses different frames to reconcile its role as an oil and gas producer with the ambition of being an important player in international climate negotiations. Lacking cost-effective domestic alternatives, Norway has favored carbon trading as a mitigation mechanism. By enacting climate change primarily as an economic issue, embedded in market assumptions and economic theory, Norway has managed to politically separate petroleum production from climate change, despite scientific knowledge linking them together. Norway's position is that being an oil producer is not conflicting with international climate ambitions as long as the oil is efficiently produced. In national political discourse, Norway's "clean oil" is translated as a comparative environmental and moral advantage and has become a way to inscribe the country's oil as a legitimate part of international climate solutions during a transitional phase. This cleanness emerges from the strategic use of production-based and consumption-based framings, depending on whether the argument involves the consumption of Norwegian "clean" gas abroad (as compared to "dirty" coal) or the national production of "clean" oil (as compared to "dirty" oil produced elsewhere). The argument about clean Norwegian oil and gas, also extensively discussed by Ryggvik and Kristoffersen (2015), demonstrates not only the importance of studying the interplay between framings but also the need to examine how these are enabled across scales, that is, by moving between international and national arenas. As shown in Chapter 2, storylines about clean oil are only possible by keeping oil's materiality out of the frame, as the combustion of oil, no matter where it occurs, produces the same amount of carbon dioxide (CO_2) emissions. Consequently, cleanness does not emerge from oil itself but rather from oil's interaction with technology and the energy consumed during extraction and transportation.

By analyzing documents "in action" (Prior 2003) and as a political technology (Asdal 2011), I examined how Lofoten has been inscribed and circulated in different policies and knowledge processes. Special attention was given to the Barents Sea–Lofoten area management plan, as it has played a central role in the Lofoten,

Vesterålen, and Senja (LoVeSe) controversy. The mechanisms used to enact a balance between economic use and conservation of valuable areas were specifically examined by following the trajectory of documents. The analysis illustrated how the management plan, by drawing together and working on several contexts and conflicting issues, performed this dual aim. Hence, *contexting* (Asdal and Moser 2012) was used as a lens to show the ways contexts, actors, and issues were simultaneously produced. By focusing on the active production of contexts, the chapter discussed how the management plan attempted to reconcile conflicting political agendas and environmental concerns that intersect and are linked to different scales (Bartlett and Vavrus 2017).

Furthermore, during the management plan process, LoVeSe (Nordland VI, VII, and Troms II) and other areas within the plan territory acquired a status as particularly valuable and vulnerable, based on existing knowledge. Chapter 3 specifically addressed how these categories have become important means of negotiation among different stakeholders. Overall, informants did not really question the authority of the knowledge base and emphasized the importance of "having science on their side". As such, scientific knowledge clearly constitutes an important *ally* in the petroleum debate. What was negotiated, however, were the political implications and practical meanings of these categories, namely, what kinds of matters of concern (Latour 2005) emerged and were enabled due to these categorizations. The analysis showed that valuable and vulnerable were contested, open, and unsettled categories, as some informants translated them as a matter of environmental risk and gambling with rich fishing fields, while others emphasized how these categories could motivate new technological solutions, strict safety measures, and oil preparedness. Risk is thus more complex and composite than just the idea of environmental vulnerability. Informants' accounts showed that it was often entangled in political, economic, and social aspects and used by stakeholders as a fluid category. If oil activity in the waters of LoVeSe implied few job opportunities and spin-off effects for the region, the risk was perceived as high, and an impact assessment under the Petroleum Act was not necessarily welcome. However, if the oil industry's incursion would create employment and benefit the local communities, the levels of risk were perceived as lower. Some informants argued that if the government could guarantee the latter, they favored an impact assessment; in other words, it would be worth exchanging job opportunities and spin-off effects for a certain level of risk. This suggests that in the case of LoVeSe, environmental and socioeconomic risks are interrelated and weighed against each other, as many local actors and stakeholders are concerned about the lack of economic activity and job opportunities in the region on the one hand, and fear for the fisheries and Lofoten's identity and image on the other hand. Thus, the analysis demonstrated how tensions between local and national economic expectations and priorities regarding oil activity are played out in the understanding of risk in the LoVeSe controversy.

Chapter 3 also discussed the relevance of studying mapping as a performative practice in connection to the Barents Sea–Lofoten area management plan. Focus was placed on how maps become political artifacts, as they do not only impose a particular order upon the territory they are designed to govern, but they also display their potential agency through means and objectives. Maps incorporate both problems and solutions from a particular viewpoint and are, therefore, never transparent or neutral. The analysis showed that spatiotemporal zoning is employed to integrate and coordinate both the marine environment and different conflicting and overlapping sectorial interests. The zoning system is specifically used to control relations and dynamics between actors and, thereby, domesticate potential environmental risks. These zones, however, perform a counterintuitive movement as they enact an integrated territory through spatial fragmentation. This "territorial whole" is specifically accomplished through separation, a syncretic mode of ordering (Law et al. 2013) that holds things together by keeping different sectors and industries (fisheries, oil industry, environmental conservation, etc.) distributed and apart.

To empirically approach the ways in which scientific advice and knowledge gathering have become recurrent strategies in the political debate regarding Lofoten, I employed Callon's (1986) four moments of translation, presented in Chapter 1. The reason was that it enabled me to schematically analyze different stages of a rather complex process in which a large number of actors, sectors, documents, research institutes, and ministries were involved. Following policy formation and corresponding practices as translation moments organized around four stages was a way to slow down interactions and movements for the purpose of analysis. While this is not the way things happen in real-time, as things occur simultaneously, the stepwise analysis allowed me to trace associations between network actors in a more organized manner. When examining the process through problematization, interessement, enrollment, and mobilization (Callon 1986), I realized that the Ministry of the Environment, despite designing its own role as head of the steering committee whereby it coordinated the stream of knowledge and the roles of several actors and stakeholders, was not really the *obligatory passage point* (Callon 1986). During the plan process, a major concern was how to achieve consensus regarding the knowledge that was being produced, besides the political possibilities or constraints created by this knowledge. As we see in Chapter 3, scientific knowledge became a powerful political tool turned into an obligatory passage point for both sides in the petroleum debate. Opening up the Lofoten area for oil drilling or keeping the area closed are political decisions that have to pass through the current knowledge base. The analysis also revealed a bidirectional dynamic between politics and science in the management plan process, as scientific research was translated into policy and political decision-making, whereas policy requirements continuously opened up new knowledge gaps that generated more research (Knol 2010; Olsen et al. 2016). Hence, the status as a particularly valuable and vulnerable area has become an important means of political negotiation during the last parliamentary periods, since minority parties have managed to keep LoVeSe closed to the petroleum industry (and, thereby, extend the temporary oil moratorium) in order to secure the government political support. This indicates that in socioenvironmental conflicts, the boundaries of a vulnerable nature often become important political negotiation strategies (Asdal 2011).

Hence, following the management plan's trajectory provided important insights into the ways science and politics interacted and collaborated in hybrid practices.

Chapter 4 addressed the importance of studying territorial disputes in the Ecuadorian Amazon as the outcome of spatial dynamics rooted in the underground petroleum reservoirs. Emphasis was placed on understanding the territorial dynamics between horizontal and vertical territories. An important strategy consisted of zooming in on specific interactions and associations between the underground and the surface and, thereby, illustrates through empirical data how spatial organization above the surface – "above-ground-matters" – are syncretically assembled due to a series of conflicting and overlapping political agendas. The underground was investigated as a specific kind of territory that coordinates interactions and relations between humans and nonhumans above the surface. The inquiry thus contributes to the study of the underground, an emerging research field, which has been a neglected geopolitical space (Valdivia 2015) in academic writing and within the political ecology/political economy tradition (Bebbington 2012; Bebbington and Bury 2013; Valdivia 2015). I will discuss this point further in the next section.

In Chapter 4, three ordering logics that coordinated the Amazon crude were identified: *developmentalism, destruction,* and *violation of rights.* In the case of *oil as development,* it displayed a pattern that recurrently turned to the past to make sense of the present. Accordingly, Ecuador's oil history was an essential part of the framing and, at the same time, became material for ordering. Based on the informants' viewpoints, *developmentalism* can be interpreted as a modernizing and homogenizing force moving between a series of dualisms: Modernization and development versus traditional ways of life; progress and infrastructure versus complex and uncertain nature, etc. The chapter argued that this drive towards *dualism* is inherent to modernity (Latour 1993; Law 1994) and illustrated how past and present versions of developmentalism or neo-developmentalism (North and Grinspun 2016) draw on notions about the state, the market, empirical knowledge, and incremental progress with ramifications back to the Enlightenment project. While this ordering mode highlights the importance of petroleum extraction (and extraction of natural resources in general) as a necessary means to address poverty, pursue social equality and development, this position has increasingly been challenged by several civil society actors and groups in recent years. Granting Pachamama/Nature constitutional rights and including *Sumak Kawsay* as a guiding principle in the 2008 Constitution reflect some of these concerns.

Furthermore, the analysis demonstrated how *oil as destruction* and *violation of rights* intersect and collaborate in their opposition towards developmentalism. One example is the road system that within the developmental logic is performed as an important infrastructure not only for the oil companies but also for the inhabitants of the Amazon region. In the past, roads were enacted as part of national development policies, as they integrated the country and connected the national oil industry with the international oil market. *Oil as destruction* and *violation of rights* clearly perform resistance through *counter-stories* in which roads are enacted as part of the oil issue since they opened up the territory to massive migration

and deforestation. The link between roads, agricultural colonization, and deforestation in the Ecuadorian Amazon has been extensively documented by several
authors (see Little 1992; Fontaine 2007; Finer et al. 2008) and points towards the
socioenvironmental complexity of the oil issue.

The mode of *destruction* gets close up to oil's materiality and risky associations
and is largely testimonial when patterning the negative effects of petroleum activity on the lives of humans and nonhumans in the rainforest. This is a mode that
orders oil and its destructive capacity from the standpoint of lived experience.
The analysis showed that this mode was largely *developmentalism* revisited, as it
worked to expose its failed strategies. Several informants emphasized that the
Amazon, despite contributing substantially to national economy and the gross
domestic product (GDP), had historically been excluded from economic growth
and was one of the country's least developed regions. Within *destruction*, the focus
is placed on the crude's material characteristics, such as toxicity and viscosity,
which become problematic in contact with the environment and fragile ecosystems. Here, risk is not perceived as a hypothetical probability of failure but as a
condition that people and nonhuman entities embody and encounter regularly.
Several storylines revolved around the inferior quality of the Ecuadorian crude (its
low American Petroleum Institute (API) gravity), especially in the case of Yasuní,
as the crude oil's density implies both technological and environmental challenges
and relatively low profitability in a scenario of depressed oil prices. Overall, this
ordering mode is inhabited by technical distrust fomented by decades of past
experiences with *predator technology*.

While *destruction* and *violation of rights* were identified as two distinct modes, they
usually cooperated in the data, as environmental devastation called for a strong
legal framework. As pointed out in Chapter 4, *violation of rights* worked on two
fronts: First, there are local lessons and ethical and legal implications of carrying
out petroleum extraction in sensitive areas such as Yasuní. Second, this ordering
mode "goes global" as it orders a variety of links and connections in a North–
South and environmental justice perspective to expose the existing asymmetries
related to the international framing of climate change. Initially, I considered
merging *destruction* and *violation of rights* due to their shared opposition regarding
developmentalism. As the analysis proceeded, however, I realized that they still
formed two distinct patterns in the empirical data. Oil as destruction was primarily concerned with ordering oil's material effects when entering in contact with
nature, technology, and humans. Violation of rights, however, operated in the
normative sphere by focusing on ethical and legal aspects of oil extraction in areas
that are biologically and culturally diverse.

Drawing on Strathern's (2004) concept of partial connections, Chapter 5 argued
that Yasuní, more than an integrated territory, constitutes a circuit of complex
connections that can best be approached as a *territory of in-betweens* due to several
projects and policies in tension. The state's attempt to integrate political disparities, overlapping zones, and policy frameworks has led to territorial *collapse* (Law
et al. 2013) and processes of resistance. Hence, the case of Yasuní-ITT is a relevant example of how "the expansion of the extractive frontier has often come

accompanied by territorialized conflict and social mobilization" (Bebbington and Bury 2013: 7).

The tracing of Yasuní-ITT from local resistance and territorial defense to institutionalization as a public policy showed that the oil moratorium had previously been positioned by environmental non-governmental organizations (NGOs) as the interface between local territorial struggles (conservation of protected areas and indigenous territories) and global climate change. Later, the idea of leaving fossil fuels in the ground was translated as the Yasuní-ITT Initiative's central thesis. Despite enrolling scientific knowledge and expert texts as *allies* in an attempt to stabilize and expand the network, the analysis of documents and interview data indicated that the Ecuadorian proposal struggled to enroll international actors as it did not fit within existing climate frameworks. The concept of overflow (Callon 1998a) was used to illustrate how the Yasuní-ITT Initiative actively worked on what was considered (by Ecuador) as limitations of mitigation mechanisms found in the Kyoto Protocol. Official documents and interviews with the key actors who designed the Yasuní model indicated that the ITT Initiative, in a rather syncretic manner, both combined and questioned elements from the Kyoto and Reducing Emissions from Deforestation and Forest Degradation (REDD) frameworks. The analysis discussed how the initiative replaced the concept of reduced emissions with *net avoided emissions* and, thereby, shifted the problem definition from emissions to the extraction of fossil fuels. By changing the problem definition, the Yasuní-ITT Initiative attempted to open the black box of climate change mitigation. The reframing of the climate issue and the movement from the atmosphere to the underground demonstrated, however, that climate change is not simply a matter of problematization but also a question of "scale-making processes" (Blok 2010: 898). Chapter 5 thus argued that the Yasuní-ITT Initiative struggled to calibrate the climate issue at a more concrete level, that of the underground, as opposed to the abstract global atmosphere.

Furthermore, the tracing of maps and reports that circulated in the Ecuadorian National Assembly illustrated the relevance of these *actants* in the process of declaring the ITT block of national interest. The conflict between constitutional principles in tension was solved by a map issued by the Ministry of Justice that moved groups of uncontacted peoples out of the territory and the oil blocks in question. The chapter's final section discussed how the Yasuní-ITT Initiative returned to civil society where it originated. When reassembled, Yasuní acquired a broader scope, which extended beyond the Amazon and socioenvironmental conflicts of the rainforest. In the hands of civil society and the Yasunidos collective, Yasuní-ITT was translated as a site of contestation and an opportunity to explore the boundaries of democratic practice and the current political system.

The comparison of the Integrated Management Plan for the Barents Sea–Lofoten area and the Management Plan for Yasuní National Park revealed several similarities and differences in the way territorial conflicts are handled. In both cases, spatial and temporal ordering are used in a rather syncretic way, as territories inscribed as vulnerable are also intersected by several sectorial interests competing for space. Chapter 6 addressed the way these conflicts are domesticated through zoning and how the underground becomes a particularly important drive in this process. The comparison showed that while changing the status of Yasuní-ITT as a protected area was a decision that took place in the National Assembly and was, therefore, more clearly political, politics also guided the Barents Sea–Lofoten area plan process. It appears, however, that in the case of Norway, the interplay between facts and values is more ambiguous due to the strong emphasis placed on knowledge-based decisions. In the case of Lofoten, this focus on knowledge has largely prevented politicians from "debating the real, value-laden and difficult central trade-off of use versus conservation" (Olsen et al. 2016: 299).

Furthermore, the comparison of Norway's International Climate and Forest Initiative (NICFI)/REDD+ and the Yasuní-ITT Initiative, which emerged as an important point of reference among informants, indicated that there were several links, connections, and overlapping interests between these two proposals. The most important difference between the two was that while the Norwegian initiative was a perfect fit with how the climate issue is framed by the United Nations Framework Convention on Climate Change (UNFCCC) and UN-REDD, that is, as a carbon-emissions problem, Ecuador's proposal overflowed these international frameworks due to the link between fossil-fuel extraction and climate change. Preventing fossil fuels from being released from the subsoil is not part of any mechanism within the international climate framework. My argument is that by shifting the focus from reduced emissions to avoided emissions at the level of the underground, the Yasuní-ITT Initiative caused friction, which Tsing describes as "the awkward, unequal, unstable, and creative qualities of interconnection across difference" (2005: 4). The analysis suggested that while Ecuador's climate initiative was clearly also a matter of national energy policies aimed at transforming the energy matrix, Norway's oil policies are *present as absence* in its climate and forest initiative. While NICFI/REDD+ are not directly part of the Lofoten controversy, Chapter 6 argued that carbon sequestration is enabling because it allows Norway to reconcile energy interests and climate ambitions, a view which has also been suggested in recent literature (see Røttereng 2018).

Situating the inquiry: Contributions and limitations

This book explores political processes and specific challenges related to non-extraction policies and the transition towards a post-carbon society. It can be further situated within what has been characterized as an emerging field within studies of petroleum politics and conflicts, namely that of the underground. The subsoil as a "geopolitical space" (Valdivia 2015), with its corresponding struggles and controversies, has largely been overlooked in academic scholarship (Bebbington 2012; Bebbington and Bury 2013). While arriving late to political ecology, Bebbington (2012) claims that politics of the underground and extractive industries have been around for a long time among activists and political ecologists enrolled in networks outside the academic world, especially in Latin America (Bebbington 2012: 1152–1153). Besides the political economy/political ecology approach, which focuses on how power relations and technology produce extractive "frontiers of accumulation and enclosure", another entry is rooted in science studies (Valdivia 2015: 1425). Here, we find the work of Mitchell (2009, 2011), who addresses what he calls *carbon democracy* as a set of interrelated sociotechnical arrangements and flows. Combining an STS perspective with political economy, the overarching question is how the concentration and control of energy flows and networks enable democratic practices or close them down (Mitchell 2009: 399). To examine oil's properties and related networks, Mitchell has traced a series of relations that "connected energy and politics, materials and ideas, humans and nonhumans, calculations and the objects of calculation, representations and forms of violence, and the present and the future" (2009: 422). While drawing on these views, Valdivia's (2015) work is less concerned with the politics and sociotechnical connections that make oil flow. Alternatively, she investigates the underground as a site that generates friction (Tsing 2005). Another incipient network and line of academic writing is related to what has been called the "fossil-fuel problem" (Princen et al. 2015). While not comprising a specific theory or framework, these authors are concerned with questions on how to exit the fossil-fuel era and the sustainability of post-fossil-fuel societies (Princen et al. 2015).

This book draws both on conceptualizations of the underground within STS and on perspectives related to the "fossil-fuel issue" (e.g., Martin 2015; Princen et al. 2015; Ryggvik and Kristoffersen 2015), which more clearly works on the political link between fossil fuels and climate change. My interest in the underground has been related mainly to the political and environmental controversies that it produces, with a focus on how the subterranean petroleum reservoirs actively participate in these processes through various sociomaterial practices. Accordingly, I have approached the underground not in terms of geological structures but as a kind of territory bounded by the surface and governed by political frameworks and the state's subsoil rights. In practice, these rights frequently contradict other rights and territorializations situated above the ground. By drawing on ANT, the inquiry contributes to a different kind of underground exploration beyond the conventional role of hydrocarbons in terms of "resource extractivism". Here, the underground becomes a powerful actant that engages in spatial ordering and territorial distribution above the surface and, thereby, sheds light on the materiality of politics and the spatial dimensions of power (Couldry 2008; Müller 2015) in socioenvironmental conflicts. The underground can therefore provide a useful optic to reflect on nature–society dualisms (Bebbington and Bury 2013). The inquiry highlights that oil and climate change are complexly entangled in knowledge and policy practices. Thus, the empirical analysis and comparison of Lofoten and Yasuní-ITT contribute with perspectives on how connections and links between the underground and the atmosphere are negotiated, enrolled, contested, pulled apart, and reassembled in what I have described as attempts of calibrating climate change.

While there are clear displays of power relations in my analysis, these have been approached mainly as effects of specific network relations. The inquiry brings attention to the relational dimension of power, in other words. Focus has been placed on how these relations specifically produce asymmetries between actors and entities that are densely connected and those that are only precariously interlinked in oil and climate networks. This is, I believe, one of the strengths of ANT: It provides a framework/method to trace how power is produced between heterogeneous actors that participate in its configuration. From this viewpoint, power asymmetries are generated and not something situated in specific actors or structures. This approach, however, has clear limitations when it comes to further analysis of what these power asymmetries mean or why they matter (Couldry 2008). Thus, the lack of a more direct focus on knowledge production and nature practices as power struggles has been strongly criticized by feminist scholars within STS (see Haraway 1997). Conversely, ANT provides important perspectives on how, for example, networks related to petroleum production generate power, but the approach is not necessarily helpful if we want to address this power from the perspective of "meaning-producing subjects". This has to do with ANT's extended principle of symmetry between humans and nonhumans. Since only humans can make sense of power, power asymmetries or differences will directly matter only to humans and evidently not to nonhumans, as they only participate in power production and stabilization, but not in sense-making, which therefore becomes a blind spot in my analysis. My point is that while my chosen approach can elucidate certain aspects of power, it also has analytic limitations. I acknowledge that one such limitation is that action tends to become almost an instrumental thing within ANT and, thereby, overlooks the "noninstrumental cultural meanings attributed to action by actors" (Hess 1997: 111). The limited focus given to action as culturally embedded or to cultural aspects, in general, can also make the researcher overlook his or her own role in shaping ANT accounts of specific phenomena (Müller 2015: 30). It is therefore important to make explicit our own positioning as we approach the worlds we intend to investigate.

In my own case, I grew up and live in a city in the Ecuadorian Andes. This implies that my "seeing" is from the South while attempting to relate to the North. As I explained in Chapter 1, I was familiar with the Yasuní case before I started the inquiry but had limited knowledge regarding the Lofoten case. From my perspective, Yasuní was domestic in the sense that it was close to home and a struggle that engaged many people around me, both politically and emotionally. This can, of course, have influenced the analysis. But partial perspectives (Haraway 1988; Strathern 2004) are, on the other hand, the only way we can approach our research objects and our construction of these objects: We have to see from somewhere, which means that everywhere is never an option. With reference to Haraway, Strathern (2004: 32) emphasizes that only partial perspectives carry promises of objective vision. As I see it, it is precisely our positioning and cultural situatedness that brings in a worldview; that is, a necessary perspective and a place to see from. This can be problematic and compromise our vision or it can be reflexively used as a resource to problematize, compare, and contrast phenomena. When I carried out data collection in Norway, I could draw on my fieldwork from Ecuador, which I believe added an extra filter and, thereby, complexity to the Lofoten case. What was seen as natural, common sense, or mainstream political practices were *othered* (Law 2004) by Yasuní and the struggles of the rainforest. While I could not benefit equally from the same dual vision when collecting data in Ecuador, attempts on *othering* later became an important strategy during data analysis, as I actively tried to relate and question empirical data from Ecuador based on what I had learned about the petroleum conflict in Lofoten and the Arctic. This partial and situated perspective should not overshadow the informants' points of view but can be turned into a tool of engagement and dialog and expose our own cultural *taken for grantedness* regarding what is being investigated, whereby new perspectives on both fieldwork and data analysis can creatively emerge.

Final comments

In the analytical chapters of this book, I have discussed the ambiguous character of oil and the way oil multiplies in practices that order it differently. Oil takes on different shapes and becomes different things in diverse technological, economic, environmental, and political ways of ordering that go far beyond its economic value as an energy commodity. Some of these ordering modes include each other and cooperate, whereas others work through antagonistic relationships that are organized around local resistance and the active questioning of particular ways of coordinating the relationship between petroleum production, development/ welfare, and the environment. Today, the climate issue is increasingly interfering with how petroleum resources are politically framed at different levels, a situation that adds to the multiple and uncertain character of oil. Mol and Law claim:

> Multiplicity is thus about coexistences at a single moment. To make sense of multiplicity, we need to think and write in topological ways, discovering methods for laying out a space, for laying out spaces, and for defining paths to walk through these.
>
> (2002: 8)

I strongly believe processual comparative case studies can be a way to approach these methodological challenges, as they allow the researcher to work horizontally between sites and locations while paying attention to how different events and policies are localized and distributed across scales. Drawing on lessons from ANT enables a different understanding of scale than what is common in comparative case studies, which often emphasize the bounding of the cases as a requirement for comparison. Hence, tracing connections vertically "opens up analytical opportunities" (Bartlett and Vavrus 2017: 73), as the conception of fixed and stable levels is replaced by the notion of an evolving and shifting network. I believe that this inquiry shows that constraining the research scope to the geographic locations of Lofoten and Yasuní instead of focusing on the processes as interfering networks would have limited the opportunities that actually lie in comparative methodologies. In both cases, the petroleum controversy is displayed in a series of spatial orderings that generates both overlapping and contingent territories. My argument is that these spatial dynamics constitute a highly distributed affair, which requires a glocal optic across scales. Hence, a processual approach towards comparison became my path through these transforming and fuzzy networks. Comparison, however, is never a neutral procedure, but the intentional action of bringing diverse locations or sites together with a specific purpose, namely, to see what can be achieved or learned about the world from this juxtaposition. Even if several links at different levels already existed between the Lofoten and Yasuní controversies, which originally made a comparative case study seem relevant, "the very act of comparing also constitutes a making of connections" (Strathern 2004: 51). Consequently, comparing has several important implications that I believe this book has largely demonstrated. First, by joining Yasuní and Lofoten, instead of other possible cases, certain links were simultaneously being produced and emphasized, as Lofoten brought to the fore several aspects of Yasuní, and Yasuní did the same thing for Lofoten. Second, in their encounter, they mutually relegated elements to the background of the inquiry as their interaction and dialog made other aspects much more relevant. In short, the decision to look at these cases together made visible specific similarities and differences that are not intrinsic to the cases themselves but emerged as what we can think of as relational effects when the cases were brought together by the act of comparison.

References

Asdal, K. (2011). *Politikkens Natur-Naturens Politikk*. Oslo: Universitetsforlaget.

Asdal, K. (2015). What is the issue? The transformative capacity of documents. *Distinktion: Scandinavian Journal of Social Theory*, 16(1), 74–90.

Asdal, K., & Moser, I. (2012). Experiments in context and contexting. *Science, Technology, & Human Values*, 37(4), 291–306.

Bartlett, L., & Vavrus, F. (2017). *Rethinking Case Study Research: A Comparative Approach*. London: Routledge.

Bebbington, A. (2012). Underground political ecologies: The second Annual Lecture of the cultural and political ecology specialty group of the association of American geographers. *Geoforum*, 43(6), 1152–1162.

Bebbington, A., & Bury, J. (2013). Political ecologies of the subsoil. In: A. Bebbington, & J. Bury (eds), *Subterranean Struggles: New Dynamics of Mining, Oil, and Gas in Latin America* (pp. 1–25). Austin: University of Texas Press.

Blok, A. (2010). Topologies of climate change: Actor-network theory, relational-scalar dynamics, and carbon-market overflows. *Environment and Planning D: Society and Space*, 28(5), 896–912.

Callon, M. (1986). Some elements of a sociology of translation: Domestication of the scallops and the fishermen of St. Brieuc Bay. In: J. Law (ed), *Power, Action and Belief: A New Sociology of Knowledge?* (pp. 196–233). London: Routledge & Kegan Paul.

Callon, M. (1998a). An essay on framing and overflowing: Economic externalities revisited by sociology. In: M. Callon (ed), *The Laws of the Markets* (pp. 244–269). Oxford: Blackwell Publishers.

Callon, M. (1998b). Introduction: The embeddedness of economic markets in economics. In: M. Callon (ed), *The Laws of the Markets* (pp. 1–57). Oxford: Blackwell Publishers.

Couldry, N. (2008). Actor-network theory and media: Do they connect and on what terms? In: A. Hepp, F. Krotz, S. Moores, & C. Winter (eds), *Connectivity, Networks and Flows: Conceptualizing Contemporary Communication* (pp. 93–110). Cresskill, NJ: Hampton Press.

Diemberger, H., Hastrup, K., Schaffer, S., Kennel, C. F., Sneath, D., Bravo, M., Graf, H., Hobbs, J., Davis, J., Nodari, M. L., Vassena, G., Irvine, R., Evans, C., Strathern, M., Hulme, M., Kaser, G., & Strathern, M. (2012). Communicating climate knowledge: Proxies, processes, politics. *Current Anthropology*, 53(2), 226–244.

Finer, M., Jenkins, C. N., Pimm, S. L., Keane, B., & Ross, C. (2008). Oil and gas projects in the Western Amazon: Threats to wilderness, biodiversity, and indigenous peoples. *PLOS ONE*, 1–9.

Fontaine, G. (2007). *El Precio del Petróleo: Conflictos Socio-Ambientales y Gobernabilidad en la Región Amazónica*. Quito: Flacso, IFEA, Abya-Yala.

Haraway, D. (1988). Situated knowledges: The science question in feminism and the privilege of partial perspectives. *Feminist Studies*, 14(3), 575–599.

Haraway, D. (1997). *Modest_Witness@Second_Millenium_Meets_OncoMouse*. London: Routledge.

Hess, D. J. (1997). *Science Studies: An Advanced Introduction*. New York: New York University Press.

Knol, M. (2010). Scientific advice in integrated ocean management: The process towards the Barents Sea plan. *Marine Policy*, 34(2), 252–260.

Knorr Cetina, K. (2001). Objectual practice. In: T. R. Schatzki, K. Knorr Cetina, & E. Von Savigny (eds), *The Practice Turn in Contemporary Theory* (pp. 175–188). New York: Routledge.

Latour, B. (1993). *We Have Never Been Modern*. Cambridge, MA: Harvard University Press.

Latour, B. (2005). *Reassembling the Social: An Introduction to Actor-Network Theory*. New York: Oxford University Press.

Law, J. (1994). *Organizing Modernity*. Oxford: Blackwell.

Law, J. (2003a). *Ordering and Obduracy*. Retrieved from Centre for Science Studies. Lancaster University: https://www.lancaster.ac.uk/fass/resources/sociology-online-papers/papers/law-ordering-and-obduracy.pdf.

Law, J. (2003b). *Traduction/Trahison: Notes on ANT*. Retrieved from Centre for Science Studies. Lancaster University: https://www.lancaster.ac.uk/fass/resources/sociology-online-papers/papers/law-traduction-trahison.pdf.

Law, J. (2004). *After Method: Mess in Social Science Research*. London: Routledge.

Law, J., & Moser, I. (2003). *Managing, Subjectivities and Desires*. Retrieved from Centre for Science Studies. Lancaster University: https://www.lancaster.ac.uk/fass/resources/sociology-online-papers/papers/law-moser-managing-subjectivities-desires.pdf.

Law, J., Afdal, G., Asdal, K., Lin, W.-y., Moser, I., & Singleton, V. (2013). Modes of syncretism: Notes on noncoherence. *Common Knowledge*, 20(1), 172–192.

Little, P. E. (1992). *Ecología Política de Cuyabeno: El Desarrollo No Sustentable de la Amazonía*. Quito: ILDIS, Abya-Yala.

Martin, P. L. (2015). Leaving oil under the Amazon: The Yasuní-ITT Initiative as a postpetroleum model? In: T. Princen, J. P. Manno, & P. L. Martin (eds), *Ending the Fossil Fuel Era* (pp. 119–144). Cambridge MA: MIT Press.

Mitchell, T. (2009). Carbon democracy. *Economy and Society*, 38(3), 399–432.

Mitchell, T. (2011). *Carbon Democracy: Political Power in the Age of Oil*. London: Verso.

Mol, A., & Law, J. (2002). Complexities: An introduction. In: J. Law, & A. Mol (eds), *Complexities: Social Studies of Knowledge Practices* (pp. 1–22). Durham: Duke University Press.

Müller, M. (2015). Assemblages and actor-networks: Rethinking socio-material power, politics and space. *Geography Compass*, 9(1), 27–41.

North, L. L., & Grinspun, R. (2016). Neo-extractivism and the new Latin American developmentalism: The missing piece of rural transformation. *Third World Quarterly*, 37(8), 1483–1504.

Olsen, E., Holen, S., Hoel, A. H., Buhl-Mortensen, L., & Røttingen, I. (2016). How integrated ocean governance in the Barents Sea was created by a drive for increased oil production. *Marine Policy*, 71, 293–300.

Princen, T., Manno, J. P., & Martin, P. L. (eds) (2015). *Ending the Fossil Fuel Era*. Cambridge, MA: MIT Press.

Prior, L. (2003). *Using Documents in Social Research*. London: Sage.

Røttereng, J.-K. S. (2018). When climate policy meets foreign policy: Pioneering and national interest in Norway's mitigation strategy. *Energy Research and Social Science*, 39, 216–225.

Ryggvik, H., & Kristoffersen, B. (2015). Heating up and cooling down the petrostate: The Norwegian experience. In: T. Princen, J. P. Manno, & P. L. Martin (eds), *Ending the Fossil Fuel Era* (pp. 249–275). Cambridge, MA: MIT Press.

Strathern, M. (2004). *Partial Connections*. Walnut Creek: Rowman & Littlefield Publishers.

Tsing, A. L. (2005). *Friction: An Ethnography of Global Connection*. Princeton, NJ: Princeton University Press.

Valdivia, G. (2015). Oil frictions and the subterranean geopolitics of energy regionalisms. *Environment and Planning A: Economy and Space*, 47(7), 1422–1430.

Index

Page numbers in *italic* denote figures

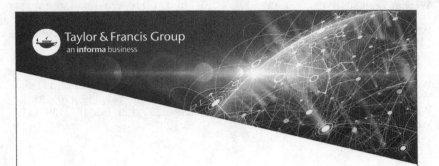

Taylor & Francis eBooks

www.taylorfrancis.com

A single destination for eBooks from Taylor & Francis
with increased functionality and an improved user
experience to meet the needs of our customers.

90,000+ eBooks of award-winning academic content in
Humanities, Social Science, Science, Technology, Engineering,
and Medical written by a global network of editors and authors.

TAYLOR & FRANCIS EBOOKS OFFERS:

A streamlined
experience for
our library
customers

A single point
of discovery
for all of our
eBook content

Improved
search and
discovery of
content at both
book and
chapter level

REQUEST A FREE TRIAL
support@taylorfrancis.com

 Routledge
Taylor & Francis Group

 CRC Press
Taylor & Francis Group

Printed in the United States
by Baker & Taylor Publisher Services

Printed in the United States
by Baker & Taylor Publisher Services